Mathematical Problem Solving

Workbook 3

Strategies for Solving Real-World Problems

Satya Pradhan

ISBN: 1541377516

ISBN-13: 978-1541377516

Contents

Acknowledgments

My sincere appreciation and thanks to the following people for their feedback and suggestions while teaching these lessons as part of an after-school math-enrichment program: Susie Bierman, Reynaldo Lorenzana, Shalini Kinger, Arun Sahoo, and Alicia Lopez. I would also like to acknowledge the help of Soumya Sahoo, Kallala Giri, and Kamalesh Parhi during the preparation of this workbook. My sincere appreciation to my wife, Nishi, for her support while I was working on this book along with my hectic, full-time job in Silicon Valley, California, and special thanks to my son, Sougat, and daughter, Sarika, for their valuable feedback on different lessons. They were my first reviewers, connecting the lessons to their classroom in school.

Introduction

Having strong problem-solving skills can make a huge difference in one's career in the modern knowledge-based economy. Problems are at the center of what we do at work every day, whether one is developing a vaccine for the winter flu, creating an antivirus for the Internet, delivering lifesaving drugs to remote villages, maximizing profits for a company, or understanding the complex structure of our universe. This means that being an effective and confident problem solver is really important to one's success. Much of that confidence comes from having a good understanding of strategy and the tools to use when approaching a problem. Therefore, it is essential for students to develop the skills and techniques for problem-solving from an early age when they are in elementary school.

Conceptual understanding, procedural and computational skills, and application of concepts to real-life problems are three pillars of mathematics education. Conceptual understanding involves knowing what to do, procedural fluency requires knowing how to do it, and problem-solving focuses on solving a wide variety of complex, real-life problems using mathematical knowledge. Mathematical skills have been taught in school by placing maximum emphasis on the understanding of math concepts and computational skills, followed by applying the concepts to real-world problems. However, the real-life problem-solving requires students to apply these concepts in the exact opposite order, starting with understanding the problem, then finding the mathematical concepts required to solve it, and finally choosing the method that best solves the problem.

Mathematical problem-solving is often taught as a way to reinforce mathematical concepts, which misses the importance of strategic thinking while solving a problem. Many research articles and books have been written emphasizing the importance of problem-solving strategies. However, the burden of teaching problem-solving strategies is left mostly to teachers and parents, who are expected to develop their own curriculum and lesson plan for the complex topic of strategy and then teach it to students.

This book presents several problem-solving strategies that teachers and parents can easily use to teach the subject. The first two chapters present the concepts of number operations and the basic problem-solving strategies listed below:
- Solving one-step problems
- Solving multistep problems
- Solving problems working backward
- Problems with too much and too little information

The concept of the unitary method is presented in chapter 3. The remaining chapters present lessons about different problem types. The objective is to teach students how to start with a problem statement, understand the problem, and then solve it with a known mathematical procedure.

Students will encounter many different problem types in their careers. We have selected the following problem types appropriate for students in third grade:
- Number problems
- Age problems
- Money problems
- Travel problem
- Work problems
- Mixture problems

Each lesson in the workbook is classified as (*), (**), or (***), depending on the level of difficulty, and each starts with a few examples showing how to solve a particular type of problem. These are followed by a number of problems of the above types.

Notes to Parents, Teachers, and Tutors

As a parent, you can use this workbook to teach problem-solving techniques to your child without any teaching experience. The first three chapters present the basic concepts and should be taught first. If students are already familiar with these concepts, these chapters can be skipped. All other chapters are independent of one another and can be taught in any order.

As a schoolteacher, you can easily integrate this workbook into the school curriculum by choosing appropriate lessons to be taught along with the curriculum.

Private tutors and after-school learning centers can use this workbook to offer special classes on mathematical problem-solving or as part of other math-enrichment programs. We suggest teaching two or three lessons a week, using the example questions provided at the beginning of each lesson and giving the other questions as homework.

Conceptual understanding of mathematical problem-solving is the main focus of this book. Therefore, we encourage students to use a calculator to solve the numerical expressions. This will allow them to take less time for numerical calculations and better focus on understanding concepts.

Answer Keys

Answer keys for all questions in this book are available online. You can download the PDF file for the answer keys at www.ilecy.com/BookAnswers

Feedback

We are always looking for feedback from students, parents, and teachers to make this book better. Please send your comments, testimonials, or suggestions for improvement to mathPS100@gmail.com.

Assessment

Note: some of the questions in the assessment may be challenging for third-grade students.

Chapter 1:

1. What operation will you use for the key words **fraction of**?
   - (a) Addition
   - (b) Subtraction
   - (c) Both (a) and (b)
   - (d) None of the above

   Answer: _____

2. Brian had 40 crayons. He gave 4 crayons to his sister and divided the remaining crayons among 4 kids. How many crayons did each kid receive?

   Answer: ____ _____
   unit

Chapter 2:

3. Andrew spent $40.00 in total while he was shopping. He spent $23.00 on a shirt, $10.00 on a wallet, and the rest on a belt. How much money did he spend on the belt?

   Answer: _____

4. Luke had some money for school supplies. He bought 3 color boxes that cost $6.00 per box. If he had $2.00 left at the end, how much money did he have at the beginning?

   Answer: _____

5. Review the question given below and choose the best answer about the available information.

   Anita had 3 pens. She bought 2 more pens and 3 baseballs. How many pens does she have in total?

   - (a) Too little information
   - (b) Too much information
   - (c) The right amount of information

   Answer: _____

Chapter 3:

6. If one shelf holds 10 books, how many shelves are needed to hold 50 books?

   Answer: ____ _____
   unit

7. 3 kids can paint a wall in 2 hours. How many kids do we need to paint the same wall in 1 hour?

   Answer: ____ _____
   unit

8. Eli and 2 of his friends (3 kids) can decorate a room in 1 hour. How many hours would it take Eli to decorate the same room by himself?

   Answer: ____ _____
   unit

Chapter 4:

9. What is the sum of the values of 5 and 3 in 453?

    Answer: _____

10. If you replace the hundreds digit by the triple of the ones digit in the number 862, what will be the new number?

    Answer: _____

11. Round 526 to the nearest ten.

    Answer: _____

12. When rounding to the nearest hundred, what is the largest number that rounds to 800?

    Answer: _____

13. Write 768 in expanded form, and find the missing number in the following math sentence:

    $768 = 700 + $ _____ $ + 8$

    Answer: _____

14. I am given a number that can be written in expanded form, as given below:

    $3,000 + 400 + 70 + 9$

    If I change 400 to 600, what will be the new number in standard form?

    Answer: _____

Chapter 5:

15. Molly is 2 years old. Amisha is presently 3 times as old as Molly. How old is Amisha?

    Answer: _____

16. The sum of Neha's and Ryan's ages is 12. What was the sum of their ages 1 year ago?

    Answer: _____

17. Alice's current age is one-half of her father's age. Her father's age is 60. How old is Alice?

    Answer: _____

18. Max is currently 16 years old. Adam is 4 years older than Max. How old is Adam?

    Answer: _____

19. The difference between Eli's and Ani's ages is 12. What was the difference between their ages 4 years ago?

    Answer: _____

20. Rob is currently 20 years old. Bill is 7 years younger than Rob. How old was Bill 2 years ago?

    Answer: _____

Chapter 6:

21. A tortoise can move 9 meters in 3 minutes. How far can it move in 1 minute?

    Answer: ____ _____
    unit

22. If it is 6:00 a.m. now, what will be the time after 4 hours?

    Answer: ____ _____
    unit

23. Michael went on a tour on March 15. He stayed there for 5 days. On which date did Michael return?
    (a) March 10
    (b) March 20
    (c) March 21
    (d) March 9

    Answer: _____

24. Helen takes 1 hour to run 5 miles. How far can she run in 3 hours?

    Answer: ____ _____
    unit

25. George went to an exhibition on June 5 and returned on June 12. How many days was George gone for the exhibition?

    Answer: ____ _____
    unit

Chapter 7:

26. 1 bottle costs $3.00, and 1` bucket costs $8.00. Ronald wants to buy 6 bottles and 5 buckets. How much money does he need?

    Answer: _____

27. Christina bought 4 sling bags for $48.00 and sold them for $55.00. How much profit did she make?

    Answer: _____

28. If 5 chairs cost $50.00, what is the cost of 1 chair?

    Answer: _____

29. Samuel bought 4 outdoor games and paid a total of $80.00. He also bought 5 tables for $60.00. What was the total cost of the games and tables?

    Answer: _____

30. Caroline wants to buy a phone that costs $89.00. She gave the cashier $100.00. How much money will she get back?

    Answer: _____

Chapter 8:

31. 3 students complete 6 assignments in 1 day. How many assignments will 1 student complete in 1 day?

    Answer: ____ _____
    unit

32. 5 workers can make 1 brick wall in 3 days. How many days will it take the workers to make 4 brick walls?

    Answer: ____ _____
    unit

33. A tap takes 1 minute to fill 5 water bottles. How many bottles will it fill in 5 minutes?

    Answer: ____ _____
    unit

34. Some workers take 9 hours to polish a floor. How long will they take to polish one-third of the floor?

    Answer: ____ _____
    unit

35. It takes 6 minutes to sing a song. How long will it take to sing 5 songs?

    Answer: ____ _____
    unit

Chapter 9:

36. Bottle 1 has 100 milliliters of juice and 100 milliliters of water. Bottle 2 has 40 milliliters of juice and 150 milliliters of water. If we mix the contents of both of the bottles, what is the total amount of the solution?

    Answer: ____ _____
    unit

37. A box has 24 apples. How many apples do you need to add to the box so that the number of apples will be 42?

    Answer: ____ _____
    unit

38. A bottle can hold 250 milliliters of liquid. The bottle has 75 milliliters of kerosene, and the rest is of an unknown liquid. What is the amount of unknown liquid in the bottle?

    Answer: ____ _____
    unit

39. Pump 1 can fill a tank in 30 minutes. Pump 2 takes 15 minutes longer than Pump 1 to fill the tank. How long will it take Pump 2 to fill the tank?

    Answer: ____ _____
    unit

40. A pipe can empty a pool in 6 hours. How long will it take to empty half of the pool?

    Answer: ____ _____
    unit

1. Mathematical Operations

1.1 Addition and Subtraction Key Words (*)

Example 1:

What is the addition or subtraction key word(s) in the following expression?

34 minus 23

Solution:

In this expression, **minus** is the subtraction key word.

Example 2:

What operation will you use for the key word **raise**?

(a) Addition
(b) Subtraction
(c) None of the above

Solution:

The key word **raise** is an **addition** key word. So the answer is (a).

Write or choose the letter of the answer.

1. What is the operation key word(s) in the following sentence?

   Mark has received an increase of $35.00 in his incentive.

   Answer: _____

2. What operation will you use for the key words **all together**?
   (a) Subtraction
   (b) Addition
   (c) Both (a) and (b)
   (d) None of the above

   Answer: _____

3. What is the operation key word(s) in the following expression?

   84 taken away from 95

   Answer: _____

4. What is the operation key word(s) in the following sentence?

   Kevin has 5 fewer toys than Gabriel.

   Answer: _____

5. What operation will you use for the key word(s) in question 4?
   (a) Subtraction
   (b) Addition
   (c) Both (a) and (b)
   (d) None of the above

   Answer: _____

6. What is the operation key word(s) in the following sentence?

   How many students are there in all?

   Answer: _____

Write or choose the letter of the answer.

7. What is the operation key word(s) in the following sentence?

    Bill had a gain of $8,500.00 in his business.

    Answer: _____

8. What operation will you use for the key word(s) in question 7?
    (a) Subtraction
    (b) Addition
    (c) Both (a) and (b)
    (d) None of the above

    Answer: _____

9. What is the operation key word(s) in the following expression?

    Remainder after 5 candies are given away

    Answer: _____

10. What operation will you use for the key word **loss**?
    (a) Subtraction
    (b) Addition
    (c) Both (a) and (b)
    (d) None of the above

    Answer: _____

11. What operation will you use for the key word **gain**?
    (a) Addition
    (b) Subtraction
    (c) Both (a) and (b)
    (d) None of the above

    Answer: _____

12. What is the operation key word(s) in the following problem?

    What is the sum of 24 and 54?

    Answer: _____

13. What operation will you use for the key words **take away**?
    (a) Subtraction
    (b) Addition
    (c) Both (a) and (b)
    (d) None of the above

    Answer: _____

14. What is the operation key word(s) in the following expression?

    Difference between 6 and 4

    Answer: _____

15. What operation will you use for the key word(s) in question 14?
    (a) Addition
    (b) Subtraction
    (c) Both (a) and (b)
    (d) None of the above

    Answer: _____

16. What operation will you use for the key words **increased by**?
    (a) Subtraction
    (b) Addition
    (c) Both (a) and (b)
    (d) None of the above

    Answer: _____

1.2 Multiplication and Division Key Words (*)

Example 1:

What is the multiplication or division key word(s) in the following sentence?

Divide 8 by 4.

Solution:

In this sentence, **divide** is the division key word.

Example 2:

What operation will you use for the key word **times**?

(a) Multiplication
(b) Division
(c) None of the above

Solution:

Times is a **multiplication** key word, so the answer is (a).

Write of choose the letter of the answer.

1. What operation will you use for the key word **triple**?
   (a) Division
   (b) Multiplication
   (c) None of the above

   Answer: _____

2. What is the operation key word(s) in the following problem?

   What is the quotient of 8 and 2?

   Answer: _____

3. What operation will you use for the key word **every**?
   (a) Multiplication
   (b) Division
   (c) None of the above

   Answer: _____

4. What is the operation key word(s) in the following problem?

   What is the product of 7 and 4?

   Answer: _____

5. What operation will you use for the key word(s) in question 4?
   (a) Division
   (b) Multiplication
   (c) None of the above

   Answer: _____

6. What operation will you use for the key word **multiply**?
   (a) Multiplication
   (b) Division
   (c) None of the above

   Answer: _____

Write or choose the letter of the answer.

7. What is the operation key word(s) in the following expression?

   One-fourth of 4

   Answer: _____

8. What operation will you use for the key word(s) in question 7?
   (a) Multiplication
   (b) Division
   (c) None of the above

   Answer: _____

9. What is the operation key word(s) in the following sentence?

   Bill traveled 62 miles in 2 hours.

   Answer: _____

10. What operation will you use for the key words **percent of**?
    (a) Division
    (b) Multiplication
    (c) None of the above

    Answer: _____

11. What operation will you use for the key words **fraction of**?
    (a) Multiplication
    (b) Division
    (c) None of the above

    Answer: _____

12. What operation will you use for the key words **divided by**?
    (a) Division
    (b) Multiplication
    (c) None of the above

    Answer: _____

13. What is the operation key word(s) in the following expression?

    40 equally divided into 5 parts

    Answer: _____

14. What operation will you use for the key word(s) in question 13?
    (a) Multiplication
    (b) Division
    (c) None of the above

    Answer: _____

15. What operation will you use for the key words **distributed equally**?
    (a) Multiplication
    (b) Division
    (c) None of the above

    Answer: _____

16. What is the operation key word(s) in the following sentence?

    Multiply 9 by 3.

    Answer: _____

1.3 Operation Key Words – 1 (*)

Example 1:

What is the operation key word(s) in the following sentence?

Birla had a gain of $150.00 in his salary.

Solution:

In this sentence, **gain** is the **addition** key word.

Example 2:

What operation will you use for the key words **equally distributed**?

(a) Subtraction
(b) Division
(c) Addition
(d) Multiplication

Solution:

The key words **equally distributed** are **division** key words, so the answer is (b).

Write or choose the letter of the answer.

1. What is the operation key word(s) in the following expression?

Difference between 98 and 45

Answer: _____

2. What is the operation key word(s) in the following expression?

Sum of 27 and 43

Answer: _____

3. What operation will you use for the key word **fewer**?
   (a) Addition
   (b) Multiplication
   (c) Division
   (d) Subtraction

Answer: _____

4. What operation will you use for the key word **remainder**?
   (a) Subtraction
   (b) Division
   (c) Addition
   (d) Multiplication

Answer: _____

5. What is the operation key word(s) in the following expression?

Product of 2, 4, and 8

Answer: _____

6. What operation will you use for the key word **loss**?
   (a) Subtraction
   (b) Multiplication
   (c) Addition
   (d) Division

Answer: _____

Write or choose the letter of the answer.

7. What is the operation key word(s) in the following sentence?

    Newton had a loss of $560.00 in a month.

    Answer: _____

8. What operation will you use for the key word **gain**?
    (a) Addition
    (b) Multiplication
    (c) Division
    (d) Subtraction

    Answer: _____

9. What operation will you use for the key word **per**?
    (a) Division
    (b) Multiplication
    (c) Addition
    (d) Subtraction

    Answer: _____

10. What is the operation key word(s) in the following sentence?

     200 pieces of chocolate are equally divided among 20 children.

     Answer: _____

11. What is the operation key word(s) in the following expression?

     87 less than 887

     Answer: _____

12. What is the operation key word(s) in the following sentence?

     20 plus 30 is 50.

     Answer: _____

13. What operation will you use for the key word **times**?
     (a) Addition
     (b) Multiplication
     (c) Subtraction
     (d) Division

     Answer: _____

14. What operation will you use for the key words **raise of**?
     (a) Multiplication
     (b) Subtraction
     (c) Addition
     (d) Division

     Answer: _____

15. What is the operation key word(s) in the following expression?

     Multiply 23 by 6

     Answer: _____

16. What operation will you use for the key word **minus**?
     (a) Subtraction
     (b) Division
     (c) Addition
     (d) Multiplication

     Answer: _____

1.4 Operation Key Words – 2 (*)

Example 1:

What is the operation key word(s) in the following sentence?

Rob is 2 times as old as his sister.

Solution:

In this sentence, **times** is the **multiplication** key word.

Example 2:

What operation will you use for the key word **remainder**?
  (a) Subtraction
  (b) Addition
  (c) Multiplication
  (d) Division

Solution:

The key word **remainder** is a **subtraction** key word. So the answer is (a).

Write or choose the letter of the answer.

1. What is the operation key word(s) in the following sentence?

   Double 22.

   Answer: _____

2. What operation will you use for the key words **take away**?
   (a) Division
   (b) Multiplication
   (c) Subtraction
   (d) Division

   Answer: _____

3. What is the operation key word(s) in the following sentence?

   10 books are equally distributed among 5 students.

   Answer: _____

4. What operation will you use for the key words **in all**?
   (a) Multiplication
   (b) Division
   (c) Subtraction
   (d) Addition

   Answer: _____

5. What is the operation key word(s) in the following sentence?

   Jack spent two-fourths of his salary while shopping.

   Answer: _____

6. What operation will you use for the key words **quotient of**?
   (a) Subtraction
   (b) Division
   (c) Addition
   (d) Multiplication

   Answer: _____

Write or choose the letter of the answer.

7. What operation will you use for the key word **more**?
   (a) Division
   (b) Multiplication
   (c) Addition
   (d) Subtraction

   Answer: _____

8. What is the operation key word(s) in the following expression?

   Three-fifths of 90

   Answer: _____

9. What is the operation key word(s) in the following sentence?

   Linky's salary increased by $50.00.

   Answer: _____

10. What operation will you use for the key words **percent of**?
   (a) Addition
   (b) Division
   (c) Multiplication
   (d) Subtraction

   Answer: _____

11. What is the operation key word(s) in the following expression?

   30% of 300

   Answer: _____

12. What is the operation key word(s) in the following sentence?

   62 minus 52 is 10.

   Answer: _____

13. What operation will you use for the key words **decreased by**?
   (a) Division
   (b) Multiplication
   (c) Addition
   (d) Subtraction

   Answer: _____

14. What operation will you use for the key word **every**?
   (a) Division
   (b) Multiplication
   (c) Subtraction
   (d) Addition

   Answer: _____

15. What is the operation key word(s) in the following expression?

   60 divided by 12

   Answer: _____

16. What operation will you use for the key words **fraction of**?
   (a) Subtraction
   (b) Multiplication
   (c) Addition
   (d) Division

   Answer: _____

1.5 Write a Math Expression (*)

Example 1:

What is the math sentence for the following problem?

1 burger costs $2.00. What is the cost of 3 burgers?

(a) $2 \div 3$
(b) 2×3
(c) $2 + 3$
(d) None of the above

Solution:

The following information is given:

Cost of 1 burger = $2.00
Number of burgers = 3

We can answer the question using the following equation:

(cost of 3 burgers)
     = (cost of 1 burger)
          × (number of burgers)

     = 2 × 3

So the answer is (b).

Example 2:

What is the operation key word(s) in the following expression?

12.75 taken away from 22

Solution:

In this expression, **taken away** are the subtraction key words.

Write or choose the letter of the answer.

1. Which math sentence will you use for the following sentence?

Ricky bought 35 crayons and 12 pencils in all.

(a) $35 - 12$
(b) $35 \div 12$
(c) 35×12
(d) $35 + 12$

Answer: _____

2. What is the operation key word(s) in the following sentence?

Two-fourths of 56 students attended the meeting.

Answer: _____

3. What is the operation key word(s) in the following expression?

Total of 56 and 76

Answer: _____

Write or choose the letter of the answer.

4. 1 pen costs $3.00. Which math sentence will you use to find the cost of 20 pens?
    (a) 3 + 20
    (b) 3 × 20
    (c) 20 ÷ 3
    (d) All of the above

    Answer: _____

5. What operation will you use for the key words **product of**?
    (a) Subtraction
    (b) Multiplication
    (c) Addition
    (d) Division

    Answer: _____

6. What is the operation key word(s) in the following expression?

    Triple 33
    (a) 33
    (b) Triple
    (c) None of the above

    Answer: _____

7. Which math sentence will you use for the following sentence?

    Vicky bought 12 books from the total of 20 books.
    (a) 12 + 20
    (b) 20 − 12
    (c) 20 ÷ 12
    (d) 20 × 12

    Answer: _____

8. What operation will you use for the key words **equally distributed**?
    (a) Addition
    (b) Division
    (c) Multiplication
    (d) Subtraction

    Answer: _____

9. What is the operation key word(s) in the following sentence?

    Lisa can read 15 newspapers in 1 day. Andy can read 3 more than Lisa.

    Answer: _____

10. What is the operation key word(s) in the following sentence?

    Nancy solved the sum of 60 and 20.

    Answer: _____

11. 1 mug costs $1.00. Which math sentence will you use to find the cost of 26 mugs?
    (a) 1.00 + 26
    (b) 26 × 1.00
    (c) 26 ÷ 1.00
    (d) All of the above

    Answer: _____

12. What is the operation key word(s) in the following expression?

    Total of 56 and 76

    Answer: _____

1.6 Write a Math Expression with Multiple Operations (*)

Example 1:

What is the math sentence for the following expression?

5 subtracted from half of 16

(a) $\left(\dfrac{1}{2} \times 16\right) - 5$

(b) $\left(\dfrac{1}{2} \div 16\right) + 5$

(c) $\left(\dfrac{1}{2} + 16\right) \div 5$

(d) $\left(\dfrac{1}{2} \times 16\right) \times 5$

Solution:

We can write the math sentence as follows:

- Half of 16 = $\dfrac{1}{2}$ of 16

  $= \dfrac{1}{2} \times 16$

- 5 subtracted from (half of 16)

  $= 5$ subtracted from $\left(\dfrac{1}{2} \times 16\right)$

  $= \left(\dfrac{1}{2} \times 16\right) - 5$

So the answer is (a).

Example 2:

What is the math sentence for the following expression?

10 added to the quotient of 20 and 4

(a) $(20 + 4) \div 10$
(b) $(20 \div 10) \times 4$
(c) $(20 \times 4) + 10$
(d) $(20 \div 4) + 10$

Solution:

We can write the math sentence as follows:

10 added to (the quotient of 20 and 4)

- The quotient of 20 and 4 = $(20 \div 4)$
- 10 added to (the quotient of 20 and 4)
  $= $ (the quotient of 20 and 4) $+ 10$
  $= (20 \div 4) + 10$

So the answer is (d).

Write or choose the letter of the answer.

1. What is the math sentence for the following expression?

   Triple of 12 divided by 4.

   (a) $(12 \times 3) - 4$
   (b) $(12 \times 3) \div 4$
   (c) $(12 \div 3) \times 4$

   Answer: _____

2. What operation will you use for the key words **percent of**?

   (a) Subtraction
   (b) Addition
   (c) Division
   (d) Multiplication

   Answer: _____

Write or choose the letter of the answer.

3. Fredrick spent $88.00 in total. He spent $32.00 on a television, $18.00 on a shirt, and the rest on a watch. How much money did he spend on the watch?

Answer: _____

4. What is the math sentence for the following expression?

10 more than 30% of 90

(a) $(0.3 \times 90) + 10$
(b) $(0.3 \times 90) \div 10$
(c) $(0.3 \div 90) - 10$
(d) $(0.3 + 90) \times 10$

Answer: _____

5. Mr. Shinha had 50 pounds of sugar for his restaurant. He used $\frac{3}{5}$ of the sugar. How many pounds of sugar did he use?

Answer: ____ _____
          unit

6. What is the math sentence for the following expression?

Sum of 20 and 15 multiplied with 3

(a) $(20 + 15) \times 3$
(b) $(20 - 15) \times 3$
(c) $(20 \times 3) + 15$
(d) $(20 \div 3) + 15$

Answer: _____

7. Nikita had 35 colored papers. She gave $\frac{2}{5}$ of the papers to her friend. How many colored papers did she have left?

Answer: ____ _____
          unit

8. What is the math sentence for the following expression?

One-fourth of 80 divided by $\frac{1}{2}$

(a) $\left(\frac{1}{4} + 80\right) \div \frac{1}{2}$

(b) $\left(\frac{1}{4} \div 80\right) \div \frac{1}{2}$

(c) $\left(\frac{1}{4} * 80\right) \div \frac{1}{2}$

(d) $\left(\frac{1}{4} * 80\right) \times 2$

Answer: _____

9. What is the math sentence for the following expression?

2.3 more than 10% of 60

(a) $(0.1 \times 60) - 2.3$
(b) $(0.1 + 60) + 2.3$
(c) $(0.1 \div 60) - 2.3$
(d) $(0.1 \times 60) + 2.3$

Answer: _____

1.7 Review of Chapter 1 (*)

Write or choose the letter of the answer.

1. What operation will you use for the key word **loss**?
    (a) Division
    (b) Subtraction
    (c) Multiplication
    (d) Addition

    Answer: _____

2. What is the operation key word(s) in the following expression?

    56 divided by 8

    Answer: _____

3. What operation will you use for the key word(s) in question 2?
    (a) Multiplication
    (b) Addition
    (c) Division
    (d) Subtraction

    Answer: _____

4. What is the math sentence for the following problem?

    Lee had 22 candies. He bought 8 more candies from a shop. How many candies did Lee have in total?

    (a) 22 ÷ 8
    (b) 22 × 8
    (c) 22 − 8
    (d) 22 + 8

    Answer: _____

5. What operation will you use for the key words **all together**?
    (a) Subtraction
    (b) Addition
    (c) Multiplication
    (d) Division

    Answer: _____

6. What is the operation key word(s) in the following sentence?

    Mr. Shah is 2 times as old as Sam.

    Answer: _____

7. What operation will you use for the key words **increased by**?
    (a) Subtraction
    (b) Division
    (c) Multiplication
    (d) Addition

    Answer: _____

8. Rohit had 59 candies. He ate 4 candies and divided the remaining candies among 5 friends. How many candies did each friend get?

    Answer: ____ _____
    unit

Write or choose the letter of the answer.

9. What is the operation key word(s) in the following problem?

What is the product of 22 and 11?

Answer: _____

10. What operation will you use for the key word(s) in question 9?
    (a) Subtraction
    (b) Division
    (c) Multiplication
    (d) Addition

Answer: _____

11. Sunil's daily wages are $10.00. He received an increase of $4.00. What are his new wages?

Which math sentence will you use to answer the problem?

    (a) 10 − 4
    (b) 10 × 4
    (c) 10 + 4
    (d) All of the above

Answer: _____

12. What operation will you use for the key words **raise of**?
    (a) Addition
    (b) Division
    (c) Subtraction
    (d) Multiplication

Answer: _____

13. What operation will you use for the keywords "$\frac{3}{4}$ of"?
    (a) Subtraction
    (b) Multiplication
    (c) Division
    (d) Addition

Answer: _____

14. What is the operation key word(s) in the following expression?

12 subtracted from the product of 4 and 5

Answer: _____

15. What is the operation key word(s) in the following sentence?

The teacher asked to find the difference between 85 and 27.

Answer: _____

16. Which math sentence will you use for the following expression?

$\frac{1}{6}$ taken away from the triple of $\frac{1}{9}$

    (a) $\left(\frac{1}{9} \times 3\right) - \frac{1}{6}$

    (b) $\left(\frac{1}{9} \div 3\right) + \frac{1}{6}$

    (c) None of the above

Answer: _____

2. Basic Problem-Solving Strategies

2.1 One-Step Problems (*)

Example 1:

If 1 basket can hold 10 pounds of fruit, how many baskets are needed to hold 40 pounds of fruit?

Solution:

The following information is given:

1 basket can hold 10 pounds of fruit.

Total pounds of fruit = 40 pounds

You can find how many baskets are needed by dividing the total pounds of fruit by the amount of fruit 1 basket can hold.

   (1 basket holds)
     = (total pounds of fruit)
       ÷ (1 basket that can hold 10 pounds of fruit)
     = 40 pounds ÷ 10 pounds

     = 4 baskets

So 4 baskets are needed to hold 40 pounds of fruit.

Example 2:

David has 3 pens. His sister gave him 2 more pens. What operation will you use to find the total number of pens?

   (a) Division
   (b) Multiplication
   (c) Subtraction
   (d) Addition

Solution:

The following information is given:

   David has 3 pens.
   His sister gave him 2 more pens.

The key word **more** is an **addition** key word. So the answer is (d).

Write or choose the letter of the answer.

1. A box has 6 toy cars. Each toy car has 4 wheels. What operation will you use to find the total number of wheels?

   (a) Division
   (b) Multiplication
   (c) Subtraction
   (d) Addition

          Answer: _____

2. If one basket can hold 8 pounds of vegetables, how many baskets do we need for 48 pounds of vegetables?

          Answer: ____ _____
                    unit

Write or choose the letter of the answer.

3. What is the operation key word (s) in the following problem?

   Andy has 10 pencils. Kevin has 3 more pencils than Andy. How many pencils does Kevin have?

   (a) More
   (b) Less
   (c) Pencils
   (d) None of the above

   Answer: _____

4. If one jar can hold 9 marbles, how many jars do we need to hold 72 marbles?

   Answer: _____ _____
   unit

5. The cost of 1 packet of cookies is $9.00. Allen bought 2 packets of cookies. How much money did he give to the cashier?

   Answer: _____

6. The cost of 1 water bottle is $2.00. Lucy bought 3 water bottles. How much money did she give to the cashier?

   Answer: _____

7. The cost of 1 perfume bottle is $17.00. Amar bought 2 perfume bottles. How much money did he give to the cashier?

   Answer: _____

8. Bill has 3 kites. His brother has 2 less kites than Bill. What operation will you use to find the total number of kites?

   (a) Addition
   (b) Subtraction
   (c) Division
   (d) Multiplication

   Answer: _____

9. John has 8 crayons. His sister gave him 12 more crayons. What operation will you use to find the total number of crayons John now has?

   Answer: _____

10. A packet has 24 chocolates. All the chocolates were divided among 6 children. What operation will you use to find the total number of chocolates each child received?

    (a) Division
    (b) Multiplication
    (c) Subtraction
    (d) Addition

    Answer: _____

2.2 Multistep Problems 1 (**)

Example 1:

Allen has to make 24 paper boats for his friends. He already made half (1/2) of them in the morning. To find the number of boats left to be made, what question do you need to answer first?

   (a) How many paper boats has Allen already made?

   (b) How many paper boats does he have to make in total?

   (c) How many paper boats does he want to make?

   (d) All of the above

Solution:

You need to find the number of paper boats he already made before you can find the number of paper boats left to be made.

So the answer is (a).

Example 2:

The cost of 1 basketball is $9.00. Max bought 3 basketballs and gave $30.00 to the cashier. How much money will the cashier return?

Solution:

   Cost of 1 basketball = $9.00

   Number of basketballs = 3

   Amount given to cashier = $30.00

• Find the cost of 3 basketballs.

   Cost of 3 basketballs = $9.00 × 3

                 = $27.00

• Find the money returned.

   Money returned = $30.00 − $27.00

             = $3.00

So the cashier will return $3.00 to Max.

Example 3:

Peter spent $74.00 in total. He spent $23.00 on headphones, $18.00 on a shirt, and the rest on a watch. How much money did he spend on the watch?

Solution:

The following information is given:

   Total money spent = $74.00

   Money spent on headphones = $23.00

   Money spent on shirt = $18.00

You can use the following steps to find the answer:

Step 1: Find the amount of money spent on the headphones and shirt.

   Money spent on the headphones and shirt

        = $23.00 + $18.00

        = $41.00

Step 2: Find the amount of money spent on the watch.

   Money spent on the watch

     = (total money spent) − (money spent on the headphones and shirt)

     = $74.00 − $41.00

     = $33.00

So Peter spent $33.00 on the watch.

Write or choose the letter of the answer.

1. Mrs. Morris bought 4 baskets that cost $6.00 each. She also bought some utensils worth $32.00. How much money did she spend in total?

   Answer: _____

2. Gary has to buy 9 gifts. He already bought one-third (1/3) of them yesterday. To find the number of gifts left to be bought, what question do you need to answer first?

   (a) How many gifts does he have to buy in total?
   (b) How many gifts does he want to buy?
   (c) How many gifts has Gary already bought?
   (d) All of the above

   Answer: _____

3. Mr. Burns bought some snacks. He paid $7.00 for biscuits and $4.00 for cakes. After paying for the snacks, he was left with $10.00. How much money did he have at the beginning?

   Answer: _____

4. Kathy spent $55.00 in total. She spent $24.00 on trousers, $14.00 on footwear, and the rest on some books. How much money did she spend on books?

   Answer: _____

5. The cost of 1 bottle of hair gel is $5.00. Shaun bought 5 bottles of hair gel and gave $30.00 to the cashier. How much money will the cashier return?

   Answer: _____

6. Simon had $28.00 to buy fruit. He spent half (1/2) of the money on oranges and the rest on apples. How much money did Simon spend on apples?

   Answer: _____

7. Victoria has to collect 6 medicine samples for her project. Sarah has to collect 2 more medicine samples than Victoria. To find the total number of medicine samples to be collected by both of them, what question do you need to answer first?

   (a) How many medicine samples has Victoria already collected?
   (b) How many medicine samples has Sarah already collected?
   (c) How many medicine samples are mandatory for the project?
   (d) None of the above

   Answer: _____

2.3 Multistep Problems 2 (**)

Example 1:

Beth bought some notebooks. She paid $8.00 for short notebooks and $9.00 for long notebooks. After paying for the notebooks, she was left with $3.00. How much money did she have at the beginning?

Solution:

The following information is given:

Amount paid for short notebooks

= $8.00

Amount paid for long notebooks

= $9.00

Amount left = $3.00

You can use the following steps to find the answer:

Step 1: Find the amount of money spent on notebooks.

Amount paid for notebooks

= amount paid for short notebooks
+ amount paid for long notebooks

= $8.00 + $9.00

= $17.00

Step 2: Find the amount of money at the beginning.

Amount at the beginning

= amount paid for notebooks
+ amount left

= $17.00 + $3.00

= $20.00

So Beth had $20.00 at the beginning.

Example 2:

Robert had $12.00 to buy flowers. He spent half (1/2) of the money on marigolds and the rest on roses. How much money did Robert spend on roses?

Solution:

The following information is given:

Amount Robert had = $12.00

Amount spent on marigolds

= half of the amount Robert had

You can use the following steps to find the answer:

Step 1: Find the amount of money spent on marigolds.

Money spent on marigolds

= half of the amount Robert had

= half of $12.00

= $12.00 ÷ 2 = $6.00

Step 2: Find the amount of money spent on roses.

Money spent on roses

= (amount Robert had)
− (money spent on marigolds)

= $12.00 − $6.00

= $6.00

So Robert spent $6.00 on roses.

Write or choose the letter of the answer.

1. The cost of 1 packet of pencils is $3.00. Scott bought 4 packets of pencils and gave $15.00 to the cashier. How much money will the cashier return?

   Answer: _____

2. Noah had $25.00 to buy candies. He spent $10.00 on orange candies and the rest on mango candies. How much money did Noah spend on mango candies?

   Answer: _____

3. Mr. Burns bought some snacks. He paid $7.00 for biscuits and $4.00 for cakes. After paying for the snacks, he was left with $10.00. How much money did he have at the beginning?

   Answer: _____

4. The cost of 1 packet of biscuits is $5.00. Mrs. Fuller bought 9 packets of biscuits and gave $50.00 to the cashier. How much money will the cashier return?

   Answer: _____

5. Brittany bought 4 skirts that cost $9.00 each. She also bought some tops worth $38.00. How much money did she spend in total?

   Answer: _____

6. A toy plane costs $9.00. Bob bought 7 toy planes and gave $65.00 to the cashier.

   What question do you need to answer first to find how much money the cashier will return?

   (a) How much did 1 toy plane cost?
   (b) What is the cost of 7 toy planes?
   (c) How many toy planes are there in the store?
   (d) All of the above

   Answer: _____

7. Sonia bought some pens. She paid $21.00 for fountain pens and $17.00 for ballpoint pens. After paying for the pens, she was left with $2.00. How much money did she have at the beginning?

   Answer: _____

2.4 Work Backward – 1 (**)

Example 1:

Kapil ordered 3 toys for $7.00 each from a store. If the cashier returned $9.00, how much money did he give to the cashier?

Solution:

The following information is given:

Number of toys ordered = 3
Cost of 1 toy = $7.00
Money returned by cashier = $9.00

We can use the following steps to find the answer.

Step 1: Find the cost of 3 toys.

(cost of 3 toys) = $7.00 × 3
= $21.00

Step 2: Find the amount given to the cashier.

(amount given to the cashier)

= (cost of 3 toys)
+ (money returned by the cashier)

= $21.00 + $9.00 = $30.00

So Kapil gave $30.00 to the cashier.

Example 2:

Some workers cleaned 70 rooms in total in 1 week. If they cleaned an equal number of rooms every day, how many rooms did they clean in 1 day?

Solution:

The following information is given:

Total rooms cleaned = 70

You can solve this problem by working backward.

Step 1: Find how many days are in a week.

1 week = 7 days

Step 2: Find the number of rooms cleaned in 1 day.

(equal number of rooms cleaned in 1 day) = (total rooms cleaned)
÷ (number of days in 1 week)

= 70 ÷ 7 = 10 rooms

So they cleaned 10 rooms in 1 day.

Write the answer.

1. Sam spent $36.00 at a flower shop. He bought 3 flower buckets that each cost the same amount. What is the cost of each flower bucket?

Answer: _____

2. A plant was 25 inches tall on Sunday. It had grown by 5 inches from Friday to Sunday. How tall was the plant on Friday?

Answer: _____ _____
unit

Write the answer.

3. The cost of 1 keyboard is $4.00. Kamal bought 7 keyboards for his office. How much money did he pay for the keyboards?

Answer: _____

4. Alka bought some toys for her brother. She gave $10.00 to the cashier, and the cashier returned $2.00. What was the cost of the toys?

Answer: _____

5. Nikita bought 4 books for $7.00 each. If the cashier returned $2.00, how much money did she give to the cashier?

Answer: _____

6. Rahul was 38 inches tall in 2015. He had grown by 7 inches from 2010 to 2015. How tall was he in 2010?

Answer: ____ _____
unit

7. Simon bought a pizza for his sister. He gave $10.00 to the cashier, and the cashier returned $3.00. What was the cost of the pizza?

Answer: _____

8. Some kids ate 40 candies in total in 4 hours. If they ate an equal number of candies every hour, how many candies did they eat in 1 hour?

Answer: ____ _____
unit

9. Rob spent $52.00 at a mall. He bought 4 shirts that each cost the same amount. What is the cost of each shirt?

Answer: _____

10. John bought some snacks for his home. He gave $20.00 to the cashier, and the cashier returned $4.00. What was the cost of the snacks?

Answer: _____

11. Elina bought 4 handbags for $9.00 each from a store. If the cashier returned $2.00, how much money did she give to the cashier?

Answer: _____

12. The cost of 1 belt is $4.00. Kunal bought 7 belts for his friends. How much money did he pay for the belts?

Answer: _____

2.5 Work Backward – 2 (**)

Example 1:

John had some money for school supplies. He bought 3 notebooks with a cost of $2.00 per notebook. If he had $4.00 left at the end, how much money did he have at the beginning?

Solution:

The following information is given:

Cost of 1 notebook = $2.00
Number of notebooks bought = 3
Amount left with John = $4.00

We can use the following steps to find the answer:

- Find the cost of 3 notebooks.

  (cost of 3 notebooks) = $2.00 × 3
  = $6.00

- Find the amount John had at the beginning.

  (amount of money at the beginning)
  = $6.00 + $4.00 = $10.00

So John had $10.00 at the beginning.

Example 2:

Lucy spent some time on math homework. Then she spent 3 hours on science homework and 2 hours on a project. If she worked a total of 7 hours, how much time did she spend on math homework?

Solution:

The following information is given:

Time spent for science = 3 hours
Time spent for the project = 2 hours
Total time spent = 7 hours

We can use the following steps to find the answer:

- Find the time spent on science and the project.

  (time spent on science and the project)
  = 3 hours + 2 hours = 5 hours

- Find the time spent on math homework.

  (time spent on math)
  = 7 hours – 5 hours = 2 hours

So Lucy spent 2 hours on math homework.

Write the answer.

1. The cost of 1 packet of pencils is $2.00. Simon bought 8 packets of pencils. How much money did he pay for the pencils?

   Answer: _____

2. Rob bought some crayons for his brother. He gave $9.00 to the cashier, and the cashier returned $3.00. What was the cost of the crayons?

   Answer: _____

Write the answer.

3. A plant was 27 inches tall on Saturday. It had grown by 6 inches every day during the previous 3 days. How tall was it on Thursday?

Answer: _____ _____
                              unit

4. The cost of 1 packet of sweets is $8.00. Nancy bought 4 packets of sweets. How much money did she pay for the sweets?

Answer: _____

5. Nikhil spent some time studying. Then he spent 4 hours playing games and 2 hours watching TV. If he passed a total of 10 hours of time, how much time did he spend on studying?

Answer: _____ _____
                              unit

6. Alex bought a toy for his brother. He gave $8.00 to the cashier, and the cashier returned $2.00. What was the cost of the toy?

Answer: _____

7. The cost of 1 rice bag is $6.00. Mr. Hood bought 7 rice bags. How much money did he pay for the rice?

Answer: _____

8. Julie bought a sari for her mother. She gave $20.00 to the cashier, and the cashier returned $4.00. What was the cost of the sari?

Answer: _____

9. A tree was 45 feet tall in 2016. It had grown by 10 feet in the previous 4 years. How tall was it in 2012?

Answer: _____ _____
                              unit

10. Mark had some money for shopping for personal items. He bought 2 shirts that cost $7.00 per shirt. If he had $5.00 left at the end, how much money did he have at the beginning?

Answer: _____

2.6 Too Much and Too Little Information – 1 (*)

Example 1:

Review the question given below and choose the best answer about the available information.

Jack wants to buy a comic book. How much money does he need to pay for the book?

(a) Too little information
(b) Too much information
(c) The right amount of information

Solution:

The following information is given:

Number of comic books Jack wants to buy = 1

In the above question, the cost of the comic book is not given.

Without knowing the cost of the book, we cannot find the amount of money Jack needs.

So the answer is <u>too little information</u>, which is option (a).

Example 2:

Review the question given below and choose the best answer about the available information.

Disha has 6 storybooks. She bought 2 more storybooks. How many storybooks does she have in total?

(a) Too little information
(b) Too much information
(c) The right amount of information

Solution:

The following information is given:

Number of storybooks Disha has = 6
Number of storybooks Disha bought = 2

You can find the total number of storybooks by adding the number of storybooks Disha had before and the number of storybooks she bought.

Total number of storybooks Disha has
= 6 + 2 = 8 storybooks

The information in the question is sufficient to find the answer. So we have the right amount of information in this question.

So the answer is <u>the right amount of information,</u> which is option (c).

Write or choose the letter of the answer.

1. If the following question has enough information, find the answer to the question. Otherwise, write "No answer" for the answer.

Rohan bought 3 coloring books from a bookstore. How much money did he give to the cashier?

Answer: _____

2. If the following question has enough information, find the answer to the question. Otherwise, write "No answer" for the answer.

There are 20 boys and 14 girls in a class. If 8 girls went out class, how many students are left?

Answer: _____

Write or choose the letter of the answer.

3. Review the question given below and choose the best answer about the available information.

    Sam has 5 toy cars. He bought 8 more toy cars, 2 toy airplanes, and 1 storybook. How many toy cars does he have in total?
    (a) Too little information
    (b) Too much information
    (c) The right amount of information

                              Answer: _____

4. Review the question given below and choose the best answer about the available information.

    Rahul bought 20 pens, and he gave some pens to his friends. How many pens does he have left?
    (a) Too much information
    (b) Too little information
    (c) The right amount of information

                              Answer: _____

5. If the following question has enough information, find the answer to the question. Otherwise, write "No answer" for the answer.

    Sarika bought a sari from a store. How much money did she give to the cashier?

                              Answer: _____

6. Review the question given below and choose the best answer about the available information.

    Nikhil had 15 toys. He lost 5 toys. How many toys does he now have?
    (a) Too little information
    (b) Too much information
    (c) The right amount of   information

                              Answer: _____

7. If the following question has enough information, find the answer to the question. Otherwise, write "No answer" for the answer.

    There are 30 chairs and 20 tables in a room. If 12 chairs are broken, how many chairs and tables are left that can be used?

                              Answer: _____

8. Review the question given below and choose the best answer about the available information.

    Jasmine has 6 ribbons. She bought 3 more ribbons and 2 bracelets. How many ribbons does she have in total?
    (a) The right amount of information
    (b) Too little information
    (c) Too much information

                              Answer: _____

2.7 Too Much and Too Little Information – 2 (*)

Example 1:

If the following question has enough information, find the answer. Otherwise, write "No answer" for the answer.

There are 20 boys and 14 girls in a class. If 8 girls left, how many students are there in the class?

Solution:

The following information is given:
Number of boys in the class = 20
Number of girls in the class = 14
Number of girls who left the class = 8

We can find the answer in the following steps:

Step 1: Find the total number of students in the class.

Total students = number of boys + number of girls
= 20 + 14 = 34 students

Step 2: Find the remaining students in the class.

Remaining students = total students − number of girls who left
= 34 − 8 = 26 students

So there are 26 students in the class.

Example 2:

Review the question given below and choose the best answer about the available information.

George has 4 pencils. He bought 8 more pencils, 3 pens, and 1 storybook. How many pencils does he have in total?

(a) Too little information
(b) Too much information
(c) The right amount of information

Solution:

The following information is given:

Number of pencils George has = 4
Number of pencils George bought = 8
Number of pens George bought = 3
Number of storybooks George bought = 1

You can find the total pencils by adding the pencils George had before and the number of pencils he bought.

The information about the pens and the storybook is not required. We have too much information in this question.

So the answer is <u>too much information,</u> which is option (b).

Write or choose the letter of the answer.

1. If the following question has enough information, find the answer to the question. Otherwise, write "No answer" for the answer.

Nikhil bought 3 whiteners from a store. How much money did he give the cashier?

Answer: _____

2. If the following question has enough information, find the answer to the question. Otherwise, write "No answer" for the answer.

Alex has 8 red shirts and 6 white shirts. If he gave 3 red shirts to his brother, how many shirts are left?

Answer: _____

Write or choose the letter of the answer.

3. Review the question given below and choose the best answer for the information available.

> Vijay wants to buy a laptop. How much money does he need to pay for the laptop?

(a) Too little information
(b) Too much information
(c) The right amount of information

Answer: _____

4. Review the question given below and choose the best answer for the available information.

> John has 3 headsets. He bought 3 more headsets. How many headsets does he have in total?

(a) Too little information
(b) Too much information
(c) The right amount of information

Answer: _____

5. If the following question has enough information, find the answer to the question. Otherwise, write "No answer" for the answer.

> Emily bought some snacks from a store. How much money did she give to the cashier?

Answer: _____

6. Review the question given below and choose the best answer about the available information.

> Mark has 5 novels. He bought 2 more novels, 2 notebooks, and 1 magazine. How many novels does he have in total?

(a) Too little information
(b) Too much information
(c) The right amount of information

Answer: _____

7. Review the question given below and choose the best answer about the available information.

> Bob wants to buy a toy. How much does he need to pay for the toy?

(a) Too little information
(b) Too much information
(c) The right amount of information

Answer: _____

8. Review the question given below and choose the best answer about the available information.

> Amit has 40 marbles. He gave 13 marbles and 5 crayons to his sister. How many marbles does he have left?

(a) The right amount of information
(b) Too much information
(c) Too little information

Answer: _____

2.8 Review of Chapter 2 (1) (**)

Write or choose the letter of the answer.

1. Mrs. Cox had $60.00 to buy groceries. She spent half (1/2) of the money on rice and the rest on pulses. How much money did Mrs. Cox spend on pulses?

   Answer: _____

2. Mia has 12 candies. She wants to give 3 of them to her sister.

   What operation will you use to find how many candies Mia has left after giving 3 to her sister?

   (a) Addition
   (b) Division
   (c) Multiplication
   (d) Subtraction

   Answer: _____

3. Pamela has to buy 6 dresses. She has already bought half (1/2) of them. To find the number of dresses left to be bought, what question do you need to answer first?

   (a) How many dresses does Pamela have to buy in total?
   (b) How many dresses has Pamela already bought?
   (c) How many dresses does Pamela want to buy?
   (d) All of the above

   Answer: _____

4. Alex spent $96.00 at a shopping mall. He bought 4 casual shirts that each cost the same amount. What is the cost of each casual shirt?

   Answer: _____

5. If the following question has enough information, find the answer to the question. Otherwise, write "No answer" for the answer.

   There are 10 rain trees and 15 trumpet trees in a garden. If there are also 20 black cherry trees, how many trees are there in total?

   Answer: _____

6. The cost of 1 packet of biscuits is $4.00. Liam bought 7 packets of biscuits. How much money did he give to the cashier?

   Answer: _____

7. Carter bought 7 notebooks for $3.00 each from a store. If the cashier returned $4.00, how much money did Carter give to the cashier?

   Answer: _____

Write or choose the letter of the answer.

8. Mrs. Murray bought 3 plates that cost $8.00 each. She also bought some kitchen supplies worth $25.00. How much money did she spend in total?

    Answer: _____

9. Jeff bought some sweets for his friends. He gave $35.00 to the cashier, and the cashier returned $3.00. What was the cost of the sweets?

    Answer: _____

10. Review the question given below and choose the best answer about the available information.

    Eric has to solve 15 math problems. He already solved 4 of them. How many math problems are left to solve?

    (a) The right amount of information
    (b) Too little information
    (c) Too much information

    Answer: _____

11. If 1 bag can hold 8 types of fruits, how many bags do we need for 16 types of fruits?

    Answer: _____ _____
    unit

12. Review the question given below and choose the best answer about the available information.

    Mr. Carr wants to decorate his dining room. How many workers does he need to decorate the dining room?

    (a) Too much information
    (b) The right amount of information
    (c) Too little information

    Answer: _____

13. Workers needed to dig 90 holes in 9 hours. If they dug an equal number of holes every hour, how many holes did they dig in 1 hour?

    Answer: _____ _____
    unit

14. Joseph bought some books. He paid $14.00 for storybooks and $12.00 for science books. After paying for the books, he was left with $6.00. How much did he have at the beginning?

    Answer: _____

15. The cost of 1 pair of slippers is $8.00. Natalie bought pairs of 2 slippers. How much money did she give to the cashier?

    Answer: _____

2.9 Review of Chapter 2 (2) (**)

Write the answer.

1. Paul spent $69.00 in total. He spent $18.00 on fruits, $20.00 on vegetables, and the rest on clothes. How much money did he spend on clothes?

   Answer: _____

2. If the following question has enough information, find the answer to the question. Otherwise, write "No answer" for the answer.

   A chef has to prepare different food items made from vegetables. How many different types of vegetables does he need?

   Answer: _____

3. Mr. Cook bought some sports equipment for his son. He gave $45.00 to the cashier, and the cashier returned $1.00. What was the cost of the sports equipment?

   Answer: _____

4. An animal was 14 inches tall in the month of April. It grew by 3 inches between the months of February and April. How tall was the animal in the month of February?

   Answer: ____ _____
                   unit

5. What is the operation key word(s) in the following problem?

   Peter has 11 marbles. Bill has 4 fewer marbles than Peter. How many marbles does Bill have?

   Answer: _____

6. If 1 cupboard can hold 12 pairs of jeans, how many pairs of jeans can 4 cupboards hold?

   Answer: ____ _____
                   unit

7. Philip had some money for buying household supplies. He bought 4 soaps that cost $3.00 each. If he had $3.00 left at the end, how much money did he have at the beginning?

   Answer: _____

8. Emma bought 3 packets of noodles that cost $4.00 each. She also bought some spices worth $7.00. How much money did she spend in total?

   Answer: _____

Write or choose the letter of the answer.

9. Mrs. Morrison has to make 8 types of dishes. She has already made one-fourth (1/4) of them. To find the number of dishes left to be made, what question do you need to answer first?

   (a) How many dishes has Mrs. Morrison already made?

   (b) How many dishes does she have to make in total?

   (c) How many dishes does she want to make?

   (d) All of the above

Answer: _____

10. Jennifer and her friends planted 63 trees in 1 week. If they planted an equal number of trees every day, how many trees did they plant in 1 day?

Answer: ____ _____
                 unit

11. Roy bought 6 milk chocolates for $12.00. Each chocolate was the same price. What operation would you use to find the cost of 1 milk chocolate?

   (a) Division

   (b) Addition

   (c) Multiplication

   (d) Subtraction

Answer: _____

12. Xavier ordered 2 sandwiches for $6.00 each from a store. If the cashier returned $3.00, how much money did he give to the cashier?

Answer: _____

13. Review the question given below and choose the best answer about the available information.

Raphael has 8 basketball cards and 5 volleyball cards. Vivian has 11 basketball cards, 10 volleyball cards, and 8 football cards. How many basketball cards are there in total?

   (a) Too much information

   (b) Too little information

   (c) The right amount of information

Answer: _____

14. What is the operation key word(s) in the following problem?

Grace has 8 dolls. Casey has 1 more doll than Grace. How many dolls does Casey have?

Answer: _____

15. Sabrina had $18.00 to buy study materials. She spent half (1/2) of the money on books and the rest on notebooks. How much money did Sabrina spend on notebooks?

Answer: _____

3. Unitary Method

3.1 Unitary Method - Direct Proportions (**)

Example 1:

If 2 taxis can carry 6 passengers, how many passengers can 1 taxi carry?

Solution:

We can solve this problem as follows:

Number of passengers 2 taxis can carry
$$= 6$$

Number of passengers 1 taxi will carry
$$= 6 \div 2 = 3$$

So 1 taxi will carry 3 passengers.

Note:

More taxis will carry more passengers.

Less taxis will carry less passengers.

Example 2:

If 1 box can have 3 dozen eggs, how many dozens of eggs can 3 boxes have?

Solution:

We can solve this problem as follows:

Dozens of eggs 1 box can have = 3

Dozens of eggs 3 boxes can have = 3 × 3
$$= 9 \text{ dozen}$$

So 3 boxes can have 9 dozen eggs.

Note:

More boxes will have more eggs.

Less boxes will have less eggs.

Write the answer.

1. If 1 jar can hold 8 marbles, how many marbles can 5 jars hold?

    Answer: ____ _____
              unit

2. If 1 dog drinks 2 liters of water in a week, how many liters of water will 5 dogs drink in a week?

    Answer: ____ _____
              unit

3. If 1 water bottle contains 1 liter of water, how many liters of water can 7 bottles contain?

    Answer: ____ _____
              unit

4. If Anil runs 2 miles every day, how many miles will he run in 15 days?

    Answer: ____ _____
              unit

Write the answer.

5. If 4 people can eat 12 pounds of rice in a week, how much rice can 1 person eat?

Answer: _____ _____
unit

6. If 1 parcel holds 2 books, how many parcels are needed to hold 8 books?

Answer: _____ _____
unit

7. If 1 cookie packet has 10 cookies, how many cookies are in 4 cookie packets?

Answer: _____ _____
unit

8. If 1 jar holds 3 liters of kerosene, how many jars are needed to hold 9 liters of kerosene?

Answer: _____ _____
unit

9. If 2 bags can hold 14 shirts, how many shirts can 1 bag hold?

Answer: _____ _____
unit

10. If 1 bird can make 1 nest in a week, how many nests can 5 birds make in a week?

Answer: _____ _____
unit

11. If 3 girls can write 18 pages, how many pages can 1 girl write?

Answer: _____ _____
unit

12. If 5 carpenters can make 10 beds in a month, how many beds can 1 carpenter make?

Answer: _____ _____
unit

13. If 1 candy packet contains 20 candies, how many candies will be in 2 candy packets?

Answer: _____ _____
unit

14. If 1 drum holds 2 buckets of water, how many drums are needed to hold 6 buckets of water?

Answer: _____ _____
unit

3.2 Unitary Method – Inverse Proportions (**)

Example 1:

2 people can mow a lawn in 1 hour. How long will it take 1 person to mow the lawn?

Solution:

We can solve this problem as follows:

- Number of hours for 2 people to mow a lawn = 1 hour

- Number of hours for 1 person to mow the lawn = 2 × 1 = 2 hours

So 1 person will take 2 hours to mow the lawn.

Note:

More people can mow a lawn in less time.

Fewer people need more time to mow a lawn.

Example 2:

1 person can prepare a report in 4 hours. How many people are needed to prepare the same report in 1 hour?

Solution:

We can solve this problem as follows:

- Number of people required to prepare a report in 4 hours = 1 person

- Number of people required to prepare the report in 1 hour

    = 4 × 1
    = 4 people

So 4 people are needed to prepare the same report in 1 hour.

Write the answer.

1. 3 students can complete a science project in 2 hours. How long will it take 1 student to complete the science project?

    Answer: _____ _____
              unit

2. 2 carpenters can make a bed in 3 days. How many carpenters do we need to make the same bed in 1 day?

    Answer: _____ _____
              unit

3. 1 dog eats a packet of food in 7 days. How many dogs will it take to eat the same amount of food in 1 day?

    Answer: _____ _____
              unit

4. 5 workers can build a wall in 1 day. How long will it take 1 worker to build the wall?

    Answer: _____ _____
              unit

Write the answer.

5. 4 people can water a garden in 1 day. How long will it take 1 person to water the garden?

    Answer: _____ _____
    unit

6. 1 cat can drink a bottle of milk in 5 days. How many cats can drink the same amount of milk in 1 day?

    Answer: _____ _____
    unit

7. 3 plumbers can set up a tank in 2 hours. How many plumbers do we need to set up the same tank in 1 hour?

    Answer: _____ _____
    unit

8. 6 students can make a project in 1 week. How long will it take 1 student to make the project?

    Answer: _____ _____
    unit

9. 1 person can make some toys in 7 hours. How many people are needed to prepare those toys in 1 hour?

    Answer: _____ _____
    unit

10. 1 rat can eat a slice of butter in 4 days. How many rats can eat the same slice of butter in 1 day?

    Answer: _____ _____
    unit

11. 3 kids can eat 1 large pizza in 1 hour. How long will it take 1 kid to eat the pizza?

    Answer: _____ _____
    unit

12. 1 worker can make a room in 10 days. How many workers are needed to make the same room in 1 day?

    Answer: _____ _____
    unit

13. 2 students can write a manuscript in 4 hours. How many students do we need to write the manuscript in 1 hour?

    Answer: _____ _____
    unit

14. 1 pipe can fill a tank in 3 hours. How many pipes are needed to fill the same tank in 1 hour?

    Answer: _____ _____
    unit

3.3 Unitary Method – Time Problems (**)

Example 1:

    1 person takes 6 days to complete a task. How many days will 2 people take to complete the task?

Solution:

You can solve this problem as follows:

Number of days 1 person takes to complete a task = 6   ← given

Number of days 2 people take to complete the task = 6 ÷ 2
        = 3 days

So 2 people will take 3 days to complete the task.

Example 2:

    George and 2 of his coworkers (3 people) can repair a vehicle in one day. How many days will George take to repair the same vehicle?

Solution:

You can solve this problem as follows:

Time 3 people take to repair a vehicle
        = 1 day
Time 1 person takes to repair the vehicle
        = 1 × 3 = 3 days

So George will take 3 days to repair the same vehicle.

Example 3:

    Simon can paint a wall in 10 hours. If 1 of his friends joins him, how many hours will it take them to paint the wall?

Solution:

You can solve this problem as follows:

Number of hours 1 person takes to paint a wall = 10 hours   ← given

Number of hours 2 people take to paint the wall = 10 ÷ 2
        = 5 hours

So it will take them 5 hours to paint the wall.

Name _____

Write the answer.

1. If 4 workers can decorate a hall in 2 hours, how many hours will 1 worker take to decorate the hall?

   Answer: ____ _____
   unit

2. Mia and Julie can make 10 dolls in 1 day. If Mia wants to make the dolls alone, how many days will she take to make all 10 dolls?

   Answer: ____ _____
   unit

3. Birla and 3 of his friends (4 people) can assemble a car in 1 day. How many days will Birla take to assemble the same car?

   Answer: ____ _____
   unit

4. Nancy can write 1 assignment in 8 minutes. If 1 of her friends joins her, how many minutes will it take them to write the assignment?

   Answer: ____ _____
   unit

5. 1 kid takes 16 minutes to eat a candy packet. How many minutes will 4 kids take to eat the candy packet?

   Answer: ____ _____
   unit

6. Mark and Luke can design a special piece of wall art in 2 days. If Mark wants to design it alone, how many days will he take to design the special wall art?

   Answer: ____ _____
   unit

7. Angela and 2 of her friends (3 people) can clean a neighborhood park in 7 hours. How many hours will Angela take to clean the same park?

   Answer: ____ _____
   unit

8. 1 person takes 6 days to complete a task. How many days will 3 people take to complete the task?

   Answer: ____ _____
   unit

9. Adam can make 1 paper boat in 6 minutes. How many minutes will he take to make 5 paper boats?

   Answer: ____ _____
   unit

10. Mahi and Jay can water a garden in 1 hour. If Mahi waters the garden alone, how many hours will she take to water the garden?

    Answer: ____ _____
    unit

3.4 Unitary Method – Work Problems (**)

Example 1:

Mr. Lopez can decorate 6 apartments in 3 days. How many apartments can he decorate in 1 day?

Solution:

Number of apartments decorated in 3 days
$$= 6$$

Number of apartments decorated in 1 day
$$= 6 \div 3 \qquad \leftarrow \text{divide by } 3$$
$$= 2 \text{ apartments}$$

So Mr. Lopez can decorate 2 apartments in 1 day.

Example 2:

Michael can write 3 pages of a report in 1 hour. How many pages can he write in 3 hours?

Solution:

Number of pages written in 1 hour = 3

Number of pages written in 3 hours
$$= 3 \times 3 \qquad \leftarrow \text{multiply } 3$$
$$= 9 \text{ pages}$$

So Michael can write 9 pages in 3 hours.

Write the answer.

1. Kevin can complete 5 pages of math assignments in 1 hour. How many pages of assignments can he finish in 2 hours?

    Answer: _____ _____
    　　　　　　　　 unit

2. Nancy can feed 8 pigeons in 2 minutes. How many pigeons can she feed in 1 minute?

    Answer: _____ _____
    　　　　　　　　 unit

3. Joseph can eat 6 candies in 6 minutes. How many candies can he eat in 1 minute?

    Answer: _____ _____
    　　　　　　　　 unit

4. Lora can plant 100 plants in 1 day. How many plants can she plant in 3 days?

    Answer: _____ _____
    　　　　　　　　 unit

Write the answer.

5. Mike walks 2 miles every day. How many miles will he walk in 1 week?

Answer: ____ _____
unit

6. Mrs. Williams can make 3 sweaters in 3 days. How many sweaters can she make in 1 day?

Answer: ____ _____
unit

7. Alan can eat 8 items in 2 hours. How many items can he eat in 1 hour?

Answer: ____ _____
unit

8. Chris can repair 9 vehicles in a week. How many vehicles can he repair in 8 weeks?

Answer: ____ _____
unit

9. Ian can make 8 paintings in 4 days. How many paintings can he make in 1 day?

Answer: ____ _____
unit

10. Colin can collect 9 coins in 3 days. How many coins can he collect in 1 day?

Answer: ____ _____
unit

11. A group of workers can clean 6 floors in 1 hour. How many floors can they clean in 6 hours?

Answer: ____ _____
unit

12. Mia can wash 2 utensils in 1 minute. How many utensils can she wash in 5 minutes?

Answer: ____ _____
unit

13. Damien can run 9 miles in 3 days. How many miles can he run in 1 day?

Answer: ____ _____
unit

14. Sarah can prepare 3 recipes in 1 hour. How many recipes can she prepare in 8 hours?

Answer: ____ _____
unit

3.5 Review of Chapter 3 (**)

Write the answer.

1. 3 workers can dig some holes in 1 day. How long will it take 1 worker to dig the holes?

    Answer: ____ _____
    unit

2. 1 person can do the shopping for kitchen appliances in 2 hours. How many people are needed to do the shopping in 1 hour?

    Answer: ____ _____
    unit

3. If 2 bikes can carry 4 passengers, how many passengers can 1 bike carry?

    Answer: ____ _____
    unit

4. If 1 rabbit eats 3 kilograms of food in 1 month, how many kilograms of food can 8 rabbits eat in 1 month?

    Answer: ____ _____
    unit

5. Julia and Nicole can prepare dinner in 1 hour. If Nicole cooks alone, how many hours will she take to prepare dinner?

    Answer: ____ _____
    unit

6. If 1 basket holds 5 pounds of vegetables, how many pounds of vegetables can 6 baskets hold?

    Answer: ____ _____
    unit

7. Jason can water his garden in 8 minutes. If 1 of his friends joins him, how many minutes will it take them to water the garden?

    Answer: ____ _____
    unit

8. 1 plumber can take 8 days to work on an apartment. How many days will 4 plumbers take to work on the apartment?

    Answer: ____ _____
    unit

9. 3 workers can clean a playground in 2 days. How many workers do we need to clean the same playground in 1 day?

    Answer: ____ _____
    unit

Write the answer.

10. If 1 parrot can collect 2 pounds of food in 1 month, how many pounds of food can 7 parrots collect in 1 month?

Answer: ____ _____
unit

11. 4 doctors can attend some patients in 2 days. How many doctors do we need to attend the same number of patients in 1 day?

Answer: ____ _____
unit

12. 1 teacher can teach a chapter in 3 hours. How many teachers are needed to teach the same chapter in 1 hour?

Answer: ____ _____
unit

13. If 1 bag holds 5 juice bottles, how many juice bottles will 8 bags hold?

Answer: ____ _____
unit

14. Maria and Angela can write 8 pages in 1 hour. If Maria writes alone, how many hours will she take to write all 8 pages?

Answer: ____ _____
unit

15. If 4 bass boats can carry 8 people, how many people can 1 bass boat carry?

Answer: ____ _____
unit

16. Bill and 3 of his coworkers (4 people) can prepare lunch for an event in 2 hours. How many hours will Bill take to prepare the same lunch?

Answer: ____ _____
unit

17. 1 student takes 6 hours to prepare study notes. How many hours will 3 students take to prepare the study notes?

Answer: ____ _____
unit

18. 5 people can complete a task in 1 day. How long will it take 1 person to complete the task?

Answer: ____ _____
unit

19. Kate can arrange all her books on a table in 9 minutes. If 2 of her friends join her, how many minutes will it take them to arrange the books?

Answer: ____ _____
unit

4. Number Problems

4.1 Place-Value Concepts - 1 (*)

Example 1:

I am the largest 2-digit number with 7 as my ones digit. What number am I?

Solution:

We can find the answer by using a place-value table.

tens	ones

- The number has 2 digits. So there should be 2 boxes.

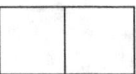

- The number is the largest number, and 7 is my ones digit. This means we have 7 in the ones place. So write 7 in the ones place.

	7

- The tens place has to be filled with the largest single number, which is 9.

9	7

So the number is 97.

Example 2:

What is the sum of the **place values** of 2 and 4 in 524?

Solution:

In the number 524:

Place value of 2 = 10
Place value of 4 = 1

Sum of the place values of 2 and 4
= 10 + 1
= 11

So the sum of the place values of 2 and 4 is 11.

Example 3:

If you replace the tens digit in 41 with the number 8, what will be the new number?

Solution:

In the number 41:

Ones digit = 1
Tens digit = 4
New number in tens digit = 8

If we replace the tens digit with the number 8, then the tens digit will be 8 instead of 4.

So the new number is 81.

Write the answer.

1. What is the value of 4 in 64?

   Answer: _____

2. If you replace the ones digit in 19 with 0, what will be the new number?

   Answer: _____

3. What is the sum of the place values of 1 and 3 in 183?

   Answer: _____

4. If you replace the ones digit in 78 with 2, what will be the new number?

   Answer: _____

5. If you replace the ones digit with the tens digit in the number 59, what is the new 2-digit number?

   Answer: _____

6. I am the largest 2-digit number with 7 as my tens digit. What number am I?

   Answer: _____

7. What is the sum of the place values of 7 and 9 in 759?

   Answer: _____

8. I am the largest 2-digit number with 3 as my ones digit. What number am I?

   Answer: _____

9. If you replace the tens digit with the ones digit in the number 24, what is the new 2-digit number?

   Answer: _____

10. What is the value of 8 in 83?

   Answer: _____

11. What is the sum of the place values of 5 and 8 in 580?

   Answer: _____

12. If you replace the tens digit in 21 by 1, what will be the new number?

   Answer: _____

4.2 Place-Value Concepts - 2 (*)

Example 1:

What is the difference between the **place values** of 8 and 2 in 821?

Solution:

In the number 821:

Place value of 8 = 100
Place value of 2 = 10

Difference between place values of 8 and 2
= 100 - 10
= 90

So the difference between the place values of 8 and 2 is 90.

Example 2:

If you replace the tens digit of 514 with 6, what will be the new number?

Solution:

In the number 514:

Ones digit = 4
Tens digit = 1
Hundreds digit = 5
New number in the tens digit = 6

If we replace the tens digit with the number 6, then the tens digit will be 6 instead of 1.

So the new number is 564.

Example 3:

What number is three times the value of 1 in 7,621?

Solution:

In the number 7,621:

Value of 1 = 1
Value of 2 = 20
Value of 6 = 600
Value of 7 = 7,000

Three times the value of 1
= 3 × (value of 1)
= 3 × 1
= 3

So 3 is three times the value of 1 in 7,621.

Example 4:

If you replace the hundreds digit with the double of the tens digit in the number 231, what will be the new number?

Solution:

In the number 231:

Ones digit = 1
Tens digit = 3
Hundreds digit = 2
Double of the tens digit = 2 × 3 = 6

If we replace the hundreds digit with the number 6, then the hundreds digit will be 6 instead of 2.

So the new number is 631.

Write the answer.

1. If you replace the tens digit of 471 with 5, what will be the new number?

    Answer: _____

2. I am the smallest 3-digit number with 2 as my ones digit. What number am I?

    Answer: _____

3. What is the difference between the values of 5 and 7 in 527?

    Answer: _____

4. If you replace the hundreds digit with the double of the tens digit in the number 147, what is the new number?

    Answer: _____

5. If you replace the hundreds digit of 342 with 5, what will be the new number?

    Answer: _____

6. I am the smallest 2-digit number with 3 as my ones digit. What number am I?

    Answer: _____

7. What is the difference between the place values of 3 and 9 in 6,394?

    Answer: _____

8. What number is two times the value of 1 in 4,310?

    Answer: _____

9. If you replace the ones digit of 114 with 2, what will be the new number?

    Answer: _____

10. I am the smallest 2-digit number with 5 as my tens digit. What number am I?

    Answer: _____

11. What number is 4 times the value of 8 in 2,608?

    Answer: _____

12. If you replace the tens digit with the double of the hundreds digit in the number 218, what is the new number?

    Answer: _____

4.3 Rounding Numbers up to Ten Thousands (**)

Example 1:

Round 5,532 to the nearest thousand.

Solution:

To round a number:

- Find the place you want to round.
- Look one place to the right.
- If the digit is less than 5, keep the rounding digit the same (do not round).
- If the digit is 5 or higher, increase the rounding digit by 1 (round up).
- Replace all the digits to the right of the rounding digit with 0.

We can find the answer by using the following steps:

Step 1: Find the digit in the thousands place. This is the place we want to round: **5**,532

Step 2: Look one place to the right: 5,**5**32 This digit (5) is **equal to** 5. So we need to **round up**.

Step 3: To round up to the nearest thousand, add 1 to the rounding number.

$(5 + 1 = 6)$

5,532 → **6**,532

Step 4: Replace all the digits to the right of the thousands place by 0.

6,**532** → 6,**000**

So 5,532 rounded to nearest the thousand is 6,000.

Example 2:

Neeraj worked at a company and was paid $331.00 in a week. About how many hundreds did he receive?

Solution:

To round a number:

- Find the place you want to round.
- Look one place to the right.
- If the digit is less than 5, keep the rounding digit the same (do not round).
- If the digit is 5 or higher, increase the rounding digit by 1 (round up).
- Replace all the digits to the right of the rounding digit with 0.

We can find the answer by using the following steps:

Step 1: Find the digit in the hundreds place. This is the place we want to round: **3**31

Step 2: Look one place to the right: 3**3**1 This digit (3) is **less than** 5. So we **do not need to** round.

Step 3: To round down to the nearest hundred, keep the hundreds-place digit the same.

331 → **3**31

Step 4: Replace all the digits to the right of the hundreds place by 0.

3**31** → 3**00**

So $331.00 rounded to the nearest hundred is $300.00.

Write the answer.

1. Round 647 to the nearest hundred.

    Answer: _____

2. Jack solved 397 math questions in a week. About how many hundreds of questions did he solve?

    Answer: _____

3. When rounding to the nearest ten, what is the smallest number that rounds to 40?

    Answer: _____

4. Round 8,745 to the nearest thousand.

    Answer: _____

5. When rounding to the nearest hundred, what is the largest number that rounds to 100?

    Answer: _____

6. Round 581 to the nearest hundred.

    Answer: _____

7. Aditya prepared 125 meals in a hotel. About how many hundreds of meals did he prepare?

    Answer: _____

8. Round 4,345 to the nearest thousand.

    Answer: _____

9. When rounding to the nearest ten, what is the smallest number that rounds to 60?

    Answer: _____

10. Arun gave 56 pieces of clothing to a charity. About how many hundreds of pieces of clothing did he give?

    Answer: _____

11. When rounding to the nearest hundred, what is the largest number that rounds to 500?

    Answer: _____

12. Round 9,641 to the nearest thousand.

    Answer: _____

4.4 Different Number Forms (*)

Example 1:

Write 924 in expanded form and find the missing number in the following math sentence:

924 = _____ + 20 + 4

Solution:

Write the numbers in the place-value table.

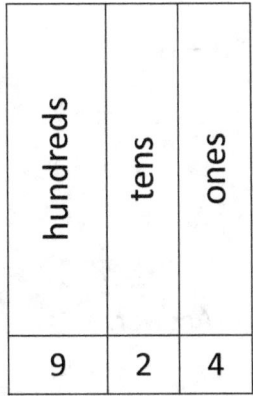

We can write the number 924 in expanded form: 9 hundreds, 2 tens, and 4 ones.

9 hundreds = 900
2 tens = 20
4 ones = 4

924 = _900_ + 20 + 4

So the missing number is 900.

Example 2:

I am given a number that is written in expanded form, as given below:

2,000 + 400 + 50 + 2

If I change 400 to 900, what will be the new number in standard form?

Solution:

First, we can write each number in expanded form:

2,000 = 2 thousands
400 = 4 hundreds
50 = 5 tens
2 = 2 ones

Then write the numbers in the place-value table.

thousands	hundreds	tens	ones
2	4	5	2

2,000 + 400 + 50 + 2 = 2,452

If I change 400 to 900, the new number will be:

2,000 + 900 + 50 + 2 = 2,952

So the new number in standard form will be 2,952.

Name _____

Write or choose the letter of the answer.

1. Write 579 in expanded form and find the missing number in the following math sentence:

   579 = _____ + 70 + 9

   Answer: _____

2. I am given a number that is written in expanded form, as given below:

   1,000 + 200 + 40 + 5

   If I change 40 to 90, what will be the new number in standard form?

   Answer: _____

3. Write 8,432 in expanded form and find the missing number in the following math sentence:

   8,432 = 8,000 + _____ + 30 + 2

   Answer: _____

4. I am given a number that is written in expanded form, as given below:

   600 + 70 + 3

   If I change 3 to 5, what will be the new number in standard form?

   Answer: _____

5. I am given a number that is written in expanded form, as given below:

   900 + 80 + 7

   If I change 900 to 500, what will be the new number in standard form?

   Answer: _____

6. Chose the answer that shows the number below in standard form:

   7,000 + 800 + 10 + 6

   (a) 7,186
   (b) 7,861
   (c) 7,816
   (d) 7,168

   Answer: _____

7. Choose the answer that shows the word form of 9,462.

   (a) Four hundred nine thousand and sixty-two
   (b) Nine thousand four hundred and twenty-six
   (c) Nine thousand four hundred and sixty-two
   (d) None of the above

   Answer: _____

8. Write 682 in expanded form and find the missing number in the following math sentence:

   682 = _____ + 80 + 2

   Answer: _____

4.5 Review of Chapter 4 (**)

Write or choose the letter of the answer.

1. What is the difference between the place values of 1 and 4 in 6,214?

    Answer: _____

2. I am given a number that is written in expanded form, as given below:

    4,000 + 900 + 10 + 9

    If I change 4,000 to 7,000, what will be the new number in standard form?

    Answer: _____

3. Kyle bought 12 toys on Sunday. About how many tens of toys did he buy?

    Answer: _____

4. Chose the answer that shows the number below in standard form:

    600 + 50 + 8

    (a) 568
    (b) 856
    (c) 658
    (d) 865

    Answer: _____

5. Round 291 to the nearest hundred.

    Answer: _____

6. I am the largest 3-digit number with 5 as my ones digit. What number am I?

    Answer: _____

7. What number is two times the value of 4 in 774?

    Answer: _____

8. Choose the answer that shows the word form of 387.
    (a) Seven hundred and thirty-eight
    (b) Three hundred and eighty-seven
    (c) Eight hundred and seventy-three
    (d) None of the above

    Answer: _____

9. What is the value of 2 in 211?

    Answer: _____

10. What is the sum of the place values of 8 and 7 in 2,847?

    Answer: _____

Write or choose the letter of the answer.

11. If you replace the tens digit of 151 with 1, what will be the new number?

Answer: _____

12. I am the smallest 2-digit number with 1 as my ones digit. What number am I?

Answer: _____

13. If you replace the hundreds digit with the ones digit in the number 917, what is the new 3-digit number?

Answer: _____

14. I am given a number that is written in expanded form, as given below:

100 + 40 + 5

If I change 100 to 600, what will be the new number in standard form?

Answer: _____

15. Write the difference between the place values of 5 and 4 in 2,540.

Answer: _____

16. When rounding to the nearest hundred, what is the largest number that rounds to 400?

Answer: _____

17. If you replace the hundreds digit by the double of the tens digit in the number 231, what is the new number?

Answer: _____

18. Write 4,341 in expanded form and find the missing number in the following math sentence:

4,341 = 4,000 + 300 + 40 + _____

Answer: _____

19. Chose the answer that shows the number below in standard form:

9,000 + 100 + 70 + 5

(a) 9,715
(b) 9,157
(c) 9,517
(d) 9,175

Answer: _____

20. What is the sum of the place values of 8 and 1 in 81?

Answer: _____

5. Age Problems

5.1 Age Problems in the Present (*)

Example 1:

Luke is currently 22 years old. Liam is 3 years older than Luke. How old is Liam?

Solution:

As given in the question:

Current age of Luke = 22 years

Current age of Liam = 3 years older than Luke

To find Liam's current age, add 3 years to Luke's current age.

(Liam's current age)

$= $ (Luke's current age) + 3

$= 22 + 3$

$= 25$ years

So Liam is 25 years old.

Example 2:

The sum of the current ages of Monica and Angela is 31. If Angela is 14 years old, how old is Monica?

Solution:

As given in the question:

Sum of Monica's and Angela's ages
= 31 years

Current age of Angela = 14 years

We can find Monica's current age as follows:

Monica's current age

$= $ (sum of Monica's and Angela's ages) − (current age of Angela)

$= 31 − 14 = 17$ years

So Monica is 17 years old.

Write the answer.

1. Disha is currently 25 years old. Lisa is 5 years older than Disha. How old is Lisa?

Answer: _____

2. Praveen is currently 19 years old. If Lora is 4 years younger than Praveen, how old is Lora?

Answer: _____

Write the answer.

3. Rohan is currently 32 years old. If Rohit is 5 years younger than Rohan, then how old is Rohit?

    Answer: _____

4. Juhi is currently 47 years old. Her mother is 15 years older than she. How old is her mother?

    Answer: _____

5. Olivia is currently 16 years old. Endy is 6 years older than Olivia. How old is Endy?

    Answer: _____

6. Rob is currently 22 years old. Bill is 10 years younger than Rob. How old is Bill?

    Answer: _____

7. The current age of Frank is 12. If Nathan is 9 years older than Frank, then how old is Nathan?

    Answer: _____

8. The sum of the current ages of Jatin and Jenny is 41. If Jatin is 21 years old, how old is Jenny?

    Answer: _____

9. Sam is currently 35 years old. If Allen is 5 years younger than Sam, how old is Allen?

    Answer: _____

10. The sum of the current ages of Neha and Rahi is 19. If Neha is 12 years old, how old is Rahi?

    Answer: _____

11. Mark is currently 38 years old. If his mother is 22 years older than he is, then how old is his mother?

    Answer: _____

12. The sum of the current ages of Kiran and her father is 65. If Kiran is 17 years old, then how old is her father?

    Answer: _____

5.2 Age Problems in the Future (**)

Example 1:

Ana will be 12 years old in 4 years. How old is she now?

Solution:

As given in the question:

Ana's age after 4 years = 12 years

We can find Ana's current age as follows:

(Ana's current age)

= (Ana's age in 4 years) − 4

= 12 − 4 = 8 years

So Ana is 8 years old now.

Example 2:

Maya is currently 9 years old. How old will she be in 7 years?

Solution:

As given in the question:

Current age of Maya = 9 years

To find Maya's age in 7 years, add 7 years to her current age.

(Maya's age in 7 years)

= (Maya's current age) + 7

= 9 + 7 = 16 years

So Maya will be 16 years old in 7 years.

Write the answer.

1. Jasmin is 20 years old. How old will she be in 2 years?

Answer: _____

2. Jennifer was 17 years old in 2015. How old will she be in 2020?

Answer: _____

3. Peter is currently 25 years old. How old will he be in 5 years?

Answer: _____

4. Simon will be 32 years old in 8 years. How old is he now?

Answer: _____

5. Carolin is 18 years old. How old will she be in 3 years?

Answer: _____

6. Akhil was 14 years old in 2015. How old will he be in 2023?

Answer: _____

Write the answer.

7. Eva is 35 years old in 2017. How old will she be in 2025?

    Answer: _____

8. Bob is currently 17 years old. How old will he be in 5 years?

    Answer: _____

9. Max will be 33 years old in 3 years. How old is he now?

    Answer: _____

10. Ayan's current age is 25. How old will he be in 10 years?

    Answer: _____

11. Angela will be 24 years old after 6 years. How old is she now?

    Answer: _____

12. Jenny is currently 29 years old. How old will she be in 6 years?

    Answer: _____

13. Helen is currently 12 years old. How old will she be in 8 years?

    Answer: _____

14. Masum will be 24 years old in 7 years. How old is she now?

    Answer: _____

15. Bill will be 28 years old after 10 years. How old is he now?

    Answer: _____

16. Rian is currently 11 years old. How old will he be in 9 years?

    Answer: _____

5.3 Age Problems in the Past (**)

Example 1:

Sam was 13 years old 2 years ago. How old is he now?

Solution:

As given in the question:

Sam's age 2 years ago = 13 years

We can find Sam's current age as follows:

(Sam's current age)

= (Sam's age 2 years ago) + 2

= 13 + 2 = 15 years

So Sam is 15 years old now.

Example 2:

Jack was 10 years old in 2012. How old was he in 2008?

Solution:

As given in the question:

Jack's age in 2012 = 10 years

We can use the following steps to answer the question:

Difference between 2012 and 2008

= 2012 − 2008

= 4 years

To find Jack's age in 2008, subtract 4 years from his age in 2012.

(Jack's age in 2008)

= (Jack's age in 2012) − 4

= 10 − 4

= 6 years

So Jack was 6 years old in 2008.

Write the answer.

1. Ryan is currently 14 years old. How old was he 4 years ago?

    Answer: _____

2. Grace was 17 years old 3 years ago. How old is Grace now?

    Answer: _____

3. Luis will be 26 years old in 6 years. How old was he 5 years ago?

    Answer: _____

4. Sofia was 31 years old in 2014. How old was she in 2006?

    Answer: _____

Write the answer.

5. Ava was 41 years old 4 years ago. How old is she now?

    Answer: _____

6. Martin was 19 years old in 2015. How old was he in 2011?

    Answer: _____

7. Jon was 23 years old in 2016. How old was he in 2008?

    Answer: _____

8. Akhil is currently 24 years old. How old was he 5 years ago?

    Answer: _____

9. Lora was 39 years old 5 years ago. How old is she now?

    Answer: _____

10. Ronnie was 18 years old in 2011. How old was he in 2007?

    Answer: _____

11. David is currently 27 years old. How old was he 10 years ago?

    Answer: _____

12. Jay's current age is 33. How old was he 6 years ago?

    Answer: _____

13. Asha was 21 years old 3 years ago. How old is she now?

    Answer: _____

14. Kyle was 18 years old in 2014. How old was he in 2011?

    Answer: _____

5.4 Review of Chapter 5 (1)(**)

Write the answer.

1. The current age of Simon is 20. If Bob is 5 years younger than Simon, how old is Bob?

    Answer: _____

2. Jack is currently 36 years old. How old was he 4 years ago?

    Answer: _____

3. Julie will be 25 years old in 5 years. How old is she now?

    Answer: _____

4. Sam will be 11 years old in 2023. How old was he in 2015?

    Answer: _____

5. Arushi is currently 29 years old. How old will she be 8 years from now?

    Answer: _____

6. The sum of Jiten's and Sam's ages is 16. If Jiten is 10 years old, how old is Sam?

    Answer: _____

7. Linda will be 8 years old in 1 year. How old was she 3 years ago?

    Answer: _____

8. Martin will be 25 years old after 8 years. How old is he now?

    Answer: _____

9. Olivia is currently 18 years old. Molly is 7 years younger than Olivia. How old is Molly now?

    Answer: _____

10. Juhi was 7 years old 6 years ago. How old is Juhi now?

    Answer: _____

Write the answer.

11. Sofia is currently 13 years old. Salini is 3 years older than Sofia. What will be Salini's age in 4 years?

Answer: _____

12. Thomas will be 18 years old in 4 years. How old is he now?

Answer: _____

13. Angela will be 20 years old in 7 years. How old was she 3 years ago?

Answer: _____

14. The sum of Mark's and Luke's ages is 27. If Mark is 12 years old, how old is Luke?

Answer: _____

15. Sarah's current age is 8, and Emily's current age is 13. What is the sum of their current ages?

Answer: _____

16. Christina is currently 24 years old. How old will she be in 6 years?

Answer: _____

17. Catherine was 12 years old in 2006. How old will she be in 2020?

Answer: _____

18. Colin was 17 years old 2 years ago. How old is Colin now?

Answer: _____

19. Rose is currently 47 years old. How old will she be 14 years from now?

Answer: _____

20. Bimal is currently 25 years old. How old was he 6 years ago?

Answer: _____

5.5 Review of Chapter 5 (2) (**)

Write the answer.

1. Jyoti is currently 24 years old. How old was she 5 years ago?

    Answer: _____

2. Disha is 5 years old. Neha is currently two times as old as Disha. How old is Neha?

    Answer: _____

3. David is currently 22. If Charles is 4 years younger than David, how old is Charles?

    Answer: _____

4. Alex is 11 years old. Nisa's age is the same as Alex's age. How old is Nisa?

    Answer: _____

5. Grace will be 17 years old after 5 years. How old is she now?

    Answer: _____

6. Anisha is 5 years old. How old will she be after 14 years?

    Answer: _____

7. Carol is presently 25 years old. Brian is 9 years older than Carol. How old is Brian?

    Answer: _____

8. Maria will be 13 years old in 4 years. How old was she 6 years ago?

    Answer: _____

9. The sum of Ricky's and Ayan's age is 35. If Ayan is 24 years old, how old is Ricky?

    Answer: _____

10. Stella is currently 15 years old. John is 5 years older than Stella. What will be John's age in 6 years?

    Answer: _____

Write the answer.

11. Bill is presently 25 years old. His mother is 25 years older than he. How old is his mother?

Answer: _____

12. Ana was 45 years old in 2014. How old was she in 2010?

Answer: _____

13. Julie's current age is 10, and Mia's current age is 16. What is the sum of their current ages?

Answer: _____

14. Ben is currently 15 years old. How old was he 9 years ago?

Answer: _____

15. Kevin is currently 7 years old. How old will he be after 7 years?

Answer: _____

16. Kavya is currently 14 years old. Jyoti is 9 years older than Kavya. How old is Jyoti now?

Answer: _____

17. Eric was 10 years old 5 years ago. How old is Eric now?

Answer: _____

18. Birla will be 25 years old after 12 years. How old is he now?

Answer: _____

19. Helen will be 18 years old in 5 years. How old was she 6 years ago?

Answer: _____

20. Adriana was 13 years old in 2013. How old was she in 2009?

Answer: _____

6. Travel Problems

6.1 Measuring Time (*)

Example 1:

If it is 8:00 p.m. now, what was the time 4 hours ago?

Solution:

The following information is given:

It is 8:00 p.m. now, and we have to find the time 4 hours ago.

Start at 8:00 p.m. and then count backward by 4 hours.

8:00 p.m. − 4 hours = 4:00 p.m.

So 4 hours ago, the time was 4:00 p.m.

Example 2:

It is 4:00 p.m. now. Justin wants to go outside after 2 hours and 30 minutes. When will he go outside?

(a) 6:30 p.m.
(b) 1:30 p.m.

Solution:

The following information is given:

It is 4:00 p.m. now, and we have to find the time after 2 hours and 30 minutes.

Start at 4:00 p.m. and then add 2 hours and 30 minutes.

4:00 p.m. + 2 hours and 30 minutes
              = 6:30 p.m.

Justin will go outside at 6:30 p.m. So the answer is (a).

Example 3:

I am a clock with three arms. My hour arm is between 6 and 7. My minute arm is at 15. If my second arm is at 30, what time do I show?

(a) 7:15:30
(b) 6:15:30
(c) 6:03:06
(d) 7:03:06

Solution:

Remember the following facts while reading a clock:

• An hour arm moves from one number to the next in one hour. So at the half hour, the hour arm points between two numbers.

• A minute arm moves a full circle in an hour (60 minutes). So it moves from one number to the next every 5 minutes.

The hour arm is between 6 and 7. So the hour is 6.

The minute arm is at 15. So the minute is 15.

The second arm is at 30. So the second is 30.

The clock will show 6:15:30. So the answer is (b).

Write or choose the letter of the answer.

1. If it is 5:15 a.m. now, what will be the time after 4 hours?

    Answer: _____

2. If it is 6:00 a.m. now, what was the time 1 hour and 30 minutes ago?

    Answer: _____

3. If it is 8:45 a.m. now, what will be the time after 1 hour?

    Answer: _____

4. I am a clock with three arms. My hour arm is between 9 and 10. My minute arm is at 45. If my second arm is at 15, what time do I show?

    (a) 10:45:15
    (b) 9:45:15
    (c) 10:15:45
    (d) 9:15:45

    Answer: _____

5. It is 4:30 p.m. now. Carol wants to go to a movie after 1 hour. When will she go to the movie?

    Answer: _____

6. I am a clock with three arms. My hour arm is between 2 and 3. My minute arm is at 40. If my second arm is at 25, what time do I show?

    (a) 3:08:25
    (b) 3:40:25
    (c) 2:08:05
    (d) 2:40:25

    Answer: _____

7. It is 6:30 p.m. now. Luke wants to go for a picnic with his friends in 1 hour and 30 minutes. When will he go to picnic?

    Answer:_____

8. If it is 8:30 p.m. now, what will be the time in 2 hours and 30 minutes?

    Answer: _____

6.2 Elapsed-Time Problems – 1 (*)

Example 1:

A party started at 7:00 p.m. and ended at 9:30 p.m. How long was the party?

Solution:

The following information is given:

The party started at 7:00 p.m. and ended at 9:30 p.m.

Start at 7:00 p.m. and count the number of hours and minutes until we reach 9:30 p.m.

First count by hours and then count by minutes.

There are 2 hours and 30 minutes between the start and end times.

So the party was 2 hours and 30 minutes.

Example 2:

Jacob returned from a 3-day sports event that ended on Saturday. On which day did the event start?

Solution:

The following information is given:

Jacob returned from a 3-day sports event. That means the event lasted for 3 days.

The event ended on Saturday. So the event had to start 2 days before Saturday.

Start with

Saturday – event ended (third day)

Friday   – second day

Thursday – first day

So the event started on Thursday.

Example 3:

If June 15 is on Monday, which day will it be on June 21?

Solution:

Write all the days in a week and their dates.

Days in a Week	Date
Monday	June 15
Tuesday	June 16
Wednesday	June 17
Thursday	June 18
Friday	June 19
Saturday	June 20
Sunday	June 21

June 21 will come after June 15.

If June 15 is on Monday, count up to June 21, which is a Sunday.

So it will be Sunday on June 21.

Method 2:

June 15 is on Monday.

June 21 will come 6 days after June 15. (June 21 – June 15 = 6 days)

So we have to count 6 days from Monday (June 15) to reach June 21.

Start with June 15 – Monday

June 16 – Tuesday

June 17 – Wednesday

June 18 – Thursday

June 19 – Friday

June 20 – Saturday

June 21 – Sunday

So it will be Sunday on June 21.

Name _____

Write the answer.

1. If February 8 is on Wednesday, which day will it be on February 12?

   Answer: _____

2. Mr. Davis returned from a 2-day summit that ended on Sunday. On which day did the summit start?

   Answer: _____

3. If April 24 is on Saturday, which day was it on April 22?

   Answer: _____

4. A teaching class started at 8:00 a.m. and ended at 8:30 p.m. How long was the class?

   Answer: ____ _____
   unit

5. John returned from a 4-day study tour that ended on Thursday. On which day did the study tour start?

   Answer: _____

6. A movie started at 5:30 p.m. and ended at 7:00 p.m. How long was the movie?

   Answer: ____ _____
   unit

7. Brian returned from a 3-day tour that ended on Tuesday. On which day did the tour start?

   Answer: _____

8. If October 13 is on Friday, which day will it be on October 16?

   Answer: _____

6.3 Elapsed-Time Problems – 2 (*)

Example 1:

Bill went for job training on August 5 and returned on August 10. How many days did Bill go for the training?

Solution:

The following information is given:

Bill went for job training on August 5 and returned on August 10.

Number of days Bill was gone for training
= the day he returned – the day he went
= August 10 – August 5
= 10 – 5 = 5 days

So Bill went for 5 days for the training.

Example 2:

A football match started at 7:00 p.m. and ended at 9:45 p.m. How long was the match?

(a) 2 hours and 45 minutes
(b) 3 hours
(c) 45 minutes
(d) 9 hours and 45 minutes

Solution:

The following information is given:

The football match started at 7:00 p.m. and ended at 9:45 p.m.

Start at 7:00 p.m. and count the number of hours until we reach 9:45 p.m.

First count by hours and then count by minutes.

There are 2 hours and 45 minutes between the start and end time.

The match was for 2 hours and 45 minutes. So the answer is (a).

Example 3:

Paul went to his uncle's house on July 3. He stayed there for 2 days. On which date did Paul return?
(a) July 1
(b) July 6
(c) July 5
(d) July 8

Solution:

The following information is given:

Paul went to his uncle's house on July 3.
Number of days he stayed = 2

He will return after 2 days from July 3.

Count 2 days after July 3.

July 3 → July 4 → July 5

Paul returned on July 5. So the answer is (c).

Write or choose the letter of the answer.

1. Pamela went to her friend's house on August 5. She stayed there for 1 day. On which date did Pamela return?

   (a) August 1
   (b) August 5
   (c) August 7
   (d) August 6

   Answer: _____

2. A seminar started at 10:00 a.m. and ended at 11:20 a.m. How long was the seminar?

   (a) 40 minutes
   (b) 20 minutes
   (c) 1 hour and 20 minutes
   (d) 1 hour and 40 minutes

   Answer: _____

3. Ann went for a trip on November 6 and returned on November 15. How many days was Ann gone for the trip?

   Answer: ____ _____
   　　　　　　 unit

4. A dance event started at 8:30 p.m. and ended at 10:50 p.m. How long was the event?

   (a) 2 hours and 20 minutes
   (b) 10 hours and 50 minutes
   (c) 8 hours and 30 minutes
   (d) 2 hours and 30 minutes

   Answer: _____

5. James went for a journey on January 12 and returned on January 23. How many days was James gone for the journey?

   Answer: ____ _____
   　　　　　　 unit

6. Mr. Nelson went abroad on December 18. He stayed there for 8 days. On which date did Mr. Nelson return?

   (a) December 27
   (b) December 8
   (c) December 10
   (d) December 26

   Answer: _____

7. Max went to a village for a survey on May 21 and returned on May 28. How many days was Max gone for the survey?

   Answer: ____ _____
   　　　　　　 unit

8. Colin went on a tour with his dad on May 14. The tour lasted for 4 days. On which date did Colin return?

   (a) May 18
   (b) May 19
   (c) May 4
   (d) May 14

   Answer: _____

Name _____

6.4 Travel Problems – 1 (*)

Example 1:

Steven can run 6 miles in 2 hours. How far can he run in 1 hour?

Solution:

The following information is given:

Distance = 6 miles

Time = 2 hours

Distance covered in 2 hours = 6 miles

Distance to be covered in 1 hour

$$= 6 \div 2 \quad \leftarrow \text{divide by 2}$$

$$= 3 \text{ miles}$$

So Steven can run 3 miles in 1 hour.

Example 2:

An ant can take 1 minute to walk 3 meters. How much distance can the ant walk in 3 minutes?

Solution:

The following information is given:

Distance = 3 meters

Time = 1 minute

Distance covered in 1 minute = 3 meters

Distance to be covered in 3 minutes

$$= 3 \times 3 \quad \leftarrow \text{multiply by 3}$$

$$= 9 \text{ meters}$$

So the ant can walk 9 meters in 3 minutes.

Write the answer.

1. Bob can take 1 minute to walk 10 meters. How much distance can he walk in 2 minutes?

Answer: ____ _____
unit

2. Mr. Parker has to travel 21 miles. How much time will it take if he travels 7 miles in 1 hour?

Answer: ____ _____
unit

3. Gloria can travel 4 kilometers in 4 minutes. How much distance can she cover in 1 minute?

Answer: ____ _____
unit

4. A bicyclist can take 1 hour to travel 10 kilometers. How much distance can he cover in 4 hours?

Answer: ____ _____
unit

Write the answer.

5. Logan can run 6 meters in 3 seconds. How much distance can he run in 1 second?

    Answer: _____ _____
    unit

6. Diego has to travel 18 miles to reach a zoo. How much time will it take if he travels 9 miles in 1 hour?

    Answer: _____ _____
    unit

7. A car traveled 4 kilometers in 4 minutes. How far can it travel in 1 minute?

    Answer: _____ _____
    unit

8. A deer has to run a distance of 80 meters in order to escape from a lion. How much time will it take if it can run 5 meters in 1 second?

    Answer: _____ _____
    unit

9. A tortoise has to swim a distance of 8 meters. How much time will it take if it swims 2 meters in 1 minute?

    Answer: _____ _____
    unit

10. A dolphin can take 1 second to swim 4 meters. How far can the dolphin swim in 6 seconds?

    Answer: _____ _____
    unit

11. A truck can take 1 minute to cover 2 miles. How far can the truck travel in 8 minutes?

    Answer: _____ _____
    unit

12. Mr. Green drives 8 meters in 2 seconds. How much distance can he drive in 1 second?

    Answer: _____ _____
    unit

6.5 Travel Problems – 2 (*)

Example 1:

    Philip can bike 80 kilometers in 2 hours. How far can he bike in 1 hour?

Solution:

The following information is given:

    Distance = 80 kilometers

    Time = 2 hours

Distance biked in 2 hours = 80 kilometers

Distance to be biked in 1 hour

$$= 80 \div 2 \quad \leftarrow \text{divide by 2}$$

$$= 40 \text{ kilometers}$$

So Philip can bike 40 kilometers in 1 hour.

Example 2:

    Kelly can walk 5 kilometers in 1 hour. How far will she travel in 3 hours?

Solution:

The following information is given:

    Distance = 5 kilometers

    Time = 1 hour

Distance covered in 1 hour = 5 kilometers

Distance to be covered in 3 hours

$$= 5 \times 3 \quad \leftarrow \text{multiply by 3}$$

$$= 15 \text{ kilometers}$$

So Kelly will cover 15 kilometers in 3 hours.

Write the answer.

1. Marcus has to swim 70 meters. How much time will he need if he can swim 2 meters in 1 second?

    Answer: ____ _____
                      unit

2. Rob can drive 99 miles in 3 hours. How far will he drive in 1 hour?

    Answer: ____ _____
                      unit

3. Liam travels 90 kilometers in 2 hours. How far can he travel in 1 hour?

    Answer: ____ _____
                      unit

4. Mark can travel 10 miles in 1 hour. How far can he travel in 5 hours?

    Answer: ____ _____
                      unit

Write the answer.

5. A rabbit can run 90 meters in 3 minutes. How far can it run in 1 minute?

Answer: ____ _____
<u>unit</u>

6. A boy has to walk 60 meters to reach his room. How long will it take him if he walks 2 meters in 1 second?

Answer: ____ _____
<u>unit</u>

7. Mr. Anderson drives 25 meters in 5 seconds. How far can he drive in 1 second?

Answer: ____ _____
<u>unit</u>

8. A shark takes 1 hour to swim 28 miles. How far can it swim in 3 hours?

Answer: ____ _____
<u>unit</u>

9. A train can travel 93 miles in 3 hours. How far will it travel in 1 hour?

Answer: ____ _____
<u>unit</u>

10. An athlete takes 1 second to run 5 meters. How far can the athlete run in 9 seconds?

Answer: ____ _____
<u>unit</u>

11. A person has to travel 60 kilometers to reach a hill station. How much time will he need if he can travel 1 kilometer in 1 minute?

Answer: ____ _____
<u>unit</u>

12. A man walks 80 meters in 2 minutes. How much distance can the man walk in 1 minute?

Answer: ____ _____
<u>unit</u>

6.6 Review of Chapter 6 (1) (*)

Write or choose the letter of the answer.

1. Laura can travel 9 kilometers in 9 minutes. How far can she travel in 1 minute?

   Answer: ____ _____

   unit

2. If it is 8:00 p.m. now, what was the time 3 hours and 30 minutes ago?

   Answer: _____

3. I am a clock with three arms. My hour arm is between 9 and 10. My minute arm is at 55. If my second arm is at 15, what time do I show?
   - (a) 10:15:55
   - (b) 9:55:15
   - (c) 10:55:15
   - (d) 9:15:55

   Answer: _____

4. A street show started at 4:30 p.m. and ended at 5:30 p.m. How long was the street show?
   - (a) 4 hours and 30 minutes
   - (b) 1 hour
   - (c) 5 hours and 30 minutes
   - (d) 30 minutes

   Answer: _____

5. If it is 7:15 a.m. now, what was the time 1 hour ago?

   Answer: _____

6. A swimmer takes 1 second to swim 4 meters. How far can the swimmer swim in 3 seconds?

   Answer: ____ _____

   unit

7. John can take 1 minute to drive 1 mile. How far can he drive in 7 minutes?

   Answer: ____ _____

   unit

8. Richard visited a place on July 13. He stayed there for 3 days. On which date did Richard return?
   - (a) July 16
   - (b) July 17
   - (c) July 13
   - (d) July 3

   Answer: _____

Write or choose the letter of the answer.

9. Bill went to a food festival on August 20 and returned on August 22. How many days was Bill gone for the festival?

    Answer: _____ _____
    　　　　　　　　unit

10. If it is 1:15 p.m. now, what will be the time after 6 hours?

    Answer: _____

11. It is 6:30 p.m. now. Edward wants to go to dinner in 2 hours. When will he go?

    Answer: _____

12. Mr. Baker went to his farmhouse on January 26. He stayed there for 4 days. On which date did he return?
    (a) January 1
    (b) January 26
    (c) January 29
    (d) January 30

    Answer: _____

13. A man has to walk 12 kilometers. How much time will he need if he can walk 6 kilometers in 1 hour?

    Answer: _____ _____
    　　　　　　　　unit

14. If October 17 is on Tuesday, which day will it be on October 21?

    Answer: _____

15. Shaun has to drive 27 miles. How much time will it take if he can travel 9 miles in 1 hour?

    Answer: _____ _____
    　　　　　　　　unit

16. Mr. Owens went for an official tour on April 2 and returned on April 7. How many days was he gone for the tour?

    Answer: _____ _____
    　　　　　　　　unit

6.7 Review of Chapter 6 (2) (*)

Write or choose the letter of the answer.

1. If December 20 is on Thursday, which day will it be on December 22?

   Answer: _____

2. Mr. Collins went for a family outing on May 9 and returned on May 16. How many days was Mr. Collins gone for the outing?

   Answer: _____ _____

   unit

3. A group of colleagues went on an excursion on July 18. They roamed around for 7 days. On which day did they return?
   (a) July 18
   (b) July 20
   (c) July 25
   (d) July 26

   Answer: _____

4. Scott returned from a 3-day science seminar that ended on Friday. On which day did the seminar start?

   Answer: _____

5. It is 7:00 a.m. now. Nicole will go to school after 2 hours. When will she go?

   Answer: _____

6. Travis takes 1 hour to cover 8 miles. How far will he travel in 4 hours?

   Answer: _____ _____

   unit

7. A section of a game started at 5:00 p.m. and ended at 5:30 p.m. How long was that section of the game?
   (a) 30 minutes
   (b) 5 hours
   (c) 5 minutes
   (d) 5 hours and 30 minutes

   Answer: _____

8. I am a clock with three arms. My hour arm is between 5 and 6. My minute arm is at 20. If my second arm is at 40, what time do I show?
   (a) 6:20:40
   (b) 6:40:20
   (c) 5:40:20
   (d) 5:20:40

   Answer: _____

Write or choose the letter of the answer.

9. If it is 9:25 p.m. now, what was the time 6 hours ago?

    Answer: _____

10. An athlete has to run 90 meters. How long will it take him if he can travel 6 meters in 1 second?

    Answer: ____ _____
    unit

11. Mrs. Torres travels 87 miles in 3 hours. How far can she travel in 1 hour?

    Answer: ____ _____
    unit

12. Kim went for an exam on February 1 and returned on February 11. How many days was Kim gone for the exam?

    Answer: ____ _____
    unit

13. If it is 5:00 a.m. now, what will be the time in 5 hours and 30 minutes?

    Answer: _____

14. A bird has to travel 92 meters. How much time will it need if it can travel 4 meters in 1 second?

    Answer: ____ _____
    unit

15. I am a clock with three arms. My hour arm is between 8 and 9. My minute arm is at 15. If my second arm is at 50, what time do I show?
    (a) 8:15:50
    (b) 8:50:15
    (c) 9:50:15
    (d) 9:15:50

    Answer: _____

16. A fish can take 1 hour to swim 8 miles. How far can the fish swim in 3 hours?

    Answer: ____ _____
    unit

7. Money Problems

7.1 Shopping Problems (*)

Example 1:

    If 8 wallets cost $40.00, what is the cost of 1 wallet?

Solution:

We can solve this problem as follows:

Cost of 8 wallets = $40.00

Cost of 1 wallet = $40.00 ÷ 8 = $5.00

So the cost of 1 wallet is $5.00.

Example 2:

    Mr. Clarke bought 6 books for $2.00 each for his daughter. How much money did he spend on books?

Solution:

We can solve this problem as follows:

Cost of 1 book = $2.00

Cost of 6 books = $2.00 × 6 = $12.00

So Mr. Clarke spent $12.00 on books.

Example 3:

    Rohit bought 6 shirts for $6.00 each and 2 pairs of jeans for $9.00 each. How much money did he pay in total?

Solution:

The following information is given:

    Cost of 1 shirt = $6.00
    Number of shirts bought = 6
    Cost of 1 pair of jeans = $9.00
    Number of pairs of jeans bought = 2

We can solve this problem as follows:

- Cost of 6 shirts

    = cost of 1 shirt × number of shirts

    = $6.00 × 6

    = $36.00

- Cost of 2 pairs of jeans

    = cost of 1 pair of jeans × number of pairs of jeans

    = $9.00 × 2

    = $18.00

Total amount paid

    = cost of 6 shirts + cost of 2 pairs of jeans

    = $36.00 + $18.00 = $54.00

So Rohit paid $54.00 in total.

Write the answer.

1. If 4 buckets cost $32.00, what is the cost of 1 bucket?

    Answer: _____

2. Elina bought 3 bracelets for $3.00 each and 4 gifts for $7.00 each. How much money did she pay in total?

    Answer: _____

3. If 5 cookie packets cost $25.00, what is the cost of 1 cookie packet?

    Answer: _____

4. Mr. Smith bought 3 novels for $9.00 each for his son. How much money did he spend in total on novels?

    Answer: _____

5. Nikhil bought 2 hats for $5.00 each and 3 pairs of trousers for $7.00 each. How much money did he pay in total?

    Answer: _____

6. Amit bought 5 watches for $35.00 for his friends. What is the cost of each watch?

    Answer: _____

7. If 1 backpack costs $9.00, what is the cost of 3 backpacks?

    Answer: _____

8. Pamela bought 4 bottles for $3.00 each for her home. How much money did she spend on bottles?

    Answer: _____

9. David bought 7 books for $5.00 each and 5 notebooks for $2.00 each. How much money did he pay in total?

    Answer: _____

10. If 9 paintings cost $72.00, what is the cost of 1 painting?

    Answer: _____

7.2 Expense Planning (*)

Example 1:

Lora wants to buy a ring that costs $140.00. She paid $155.00. How much money did she get back?

Solution:

The following information is given:

Cost of a ring = $140.00
Amount paid = $155.00

We can solve this problem as follows:

Amount of money Lora got back

= amount paid − cost of the ring

= $155.00 − $140.00

= $15.00

So Lora got back $15.00.

Example 2:

Vinod wants to buy a cooler that costs $150.00. He also needs to buy a computer that costs $600.00. How much money does he need in total?

Solution:

The following information is given:

Cost of a cooler = $150.00
Cost of a computer = $600.00

We can solve this problem as follows:

Total amount of money Vinod needs

= cost of a cooler

             + cost of a computer

= $150.00 + $600.00

= $750.00

So Vinod needs $750.00 in total.

Write the answer.

1. Sonia wants to buy a sewing machine that costs $95.00. She paid $100.00. How much money did she get back?

    Answer: _____

2. A burger costs $4.00, a candy packet costs $8.00, and a set of bottles cost $12.00. What is the cost of all three items?

    Answer: _____

3. Jack wants to buy a cycle that costs $65.00. He also needs to buy a video game that costs $22.00. How much money does he need in total?

    Answer: _____

4. Somali wants to buy a pair of sandals that costs $23.00. She paid $25.00. How much money did she get back?

    Answer: _____

Write the answer.

5. A packet of sweets costs $10.00, a packet of cashew nuts costs $18.00, and an ice cream cone costs $3.00. What is the cost of all three items?

    Answer: _____

6. Mr. Ray wants to buy a television that costs $200.00. He also needs to buy a printer that costs $65.00. How much money does he need in total?

    Answer: _____

7. Simran wants to buy books that cost $17.00. She paid $20.00. How much money did she get back?

    Answer: _____

8. A pair of sunglasses costs $42.00, a wallet costs $12.00, and a jacket costs $25.00. What is the cost of all three items?

    Answer: _____

9. Jason wants to buy a family pool that costs $72.00. He paid $80.00. How much money did he get back?

    Answer: _____

10. A chair costs $27.00, a table costs $35.00, and a washing machine costs $225.00. What is the cost of all three items?

    Answer: _____

11. Nancy wants to buy a laptop that costs $350.00. She also needs to buy a mobile phone that costs $120.00. How much money does she need in total?

    Answer: _____

12. Alex wants to buy a wall clock that costs $42.00. He paid $50.00. How much money did he get back?

    Answer: _____

7.3 Investment Problems (**)

Example 1:

David deposited an amount of $800.00 in his savings account. After five years, he had a balance of $1,200.00. How much interest did he earn in five years?

Solution:

The following information is given:

Amount of money deposited = $800.00
Time period = 5 years
Balance after 5 years = $1,200.00

We can solve this problem as follows:

Interest can be calculated by subtracting the deposited amount from the balance after 5 years.

Interest = (balance after 5 years)
         − (amount of money deposited)

    = $1,200.00 - $800.00 = $400.00

So David earned $400.00 of interest in five years.

Example 2:

Mr. Williams donates $2,900.00 to a charity, $1,750.00 to a hospital, and $1,500.00 to an institute every year. How much money does he donate every year?

Solution:

The following information is given:

Amount donated to charity = $2,900.00
Amount donated to hospital = $1,750.00
Amount donated to institute = $1,500.00

We can solve this problem as follows:

Total amount donated
    = (amount donated to charity)
         + (amount donated to hospital)
         + (amount donated to institute)

    = $2,900.00 + $1,750.00 + $1,500.00
    = $6,150.00

So Mr. Williams donates $6,150.00 every year.

Write the answer.

1. Rakesh had $950.00 in his bank account. After a few days, he withdrew $653.00 for some medical expenses. How much money is left in his account?

2. Kapil invested $1,700.00 in a mutual fund, $5,000.00 in a company's stock, and $3,500.00 in construction work. How much money did he invest in total?

Answer: _____

Answer: _____

Write the answer.

3. Dev deposited an amount of $750.00 in his savings account. After three years, he had a balance of $1,840.00. How much interest did he earn in three years?

Answer: _____

4. Alice had $1,100.00 in her bank account. After a few days, she took out $420.00 for her readmission. How much money is left in her account?

Answer: _____

5. Mr. Harper donates $1,500.00 to a child labor fund, $800.00 to a school, and $650.00 to his village youth academy every year. How much money does he donate every year?

Answer: _____

6. Dipika borrowed an amount of $1,200.00 from a bank. After five years, she repaid a total of $1,650.00. How much interest did she pay in five years?

Answer: _____

7. Nikhil deposited $1,700.00 in his savings account and $2,500.00 in his checking account. How much money did he deposit in total?

Answer: _____

8. Mrs. Hill invested an amount of $2,000.00 in a company's stock. After four years, she had a balance of $2,700.00. How much interest did she earn in four years?

Answer: _____

9. Rahul had $500.00 in his bank account. After a few days, he withdrew $235.00 for some personal expenses. How much money is left in his account?

Answer: _____

10. Kyle deposited an amount of $1,850.00 in his savings account. After six years, he had a balance of $2,300.00. How much interest did he earn in six years?

Answer: _____

7.4 Pricing Problems (**)

Example 1:

1 package of sugar costs $4.00, and 1 package of salt costs $3.00. Kapil wants to buy 3 packages of sugar and 2 packages of salt. How much money does he need?

Solution:

The following information is given:

Cost of 1 package of sugar = $4.00

Cost of 1 package of salt = $3.00

Number of packages of sugar bought = 3

Number of packages of salt bought = 2

We can solve this problem as follows:

- Total cost of 3 packages of sugar
    = cost of 1 package of sugar × 3
    = $4.00 × 3 = $12.00

- Total cost of 2 packages of salt
    = cost of 1 package of salt × 2
    = $3.00 × 2 = $6.00

- Total cost of packages of sugar and salt
    = total cost of 3 packages of sugar
        + total cost of 2 packages of salt
    = $12.00 + $6.00 = $18.00

So Kapil needs $18.00 to buy 3 packages of sugar and 2 packages of salt.

Example 2:

Bill bought 8 baseballs and paid a total of $16.00. He also bought 2 bats for $30.00. What is the total cost of the baseballs and bats?

Solution:

The following information is given:

Amount paid for baseballs = $16.00

Amount paid for bats = $30.00

We can solve this problem as follows:

Total cost of baseballs and bats
    = (amount paid for baseballs)
        + (amount paid for bats)
    = $16.00 + $30.00 = $46.00

So the total cost of the baseballs and bats is $46.00.

Write the answer.

1. The cost of 1 pizza is $6.00. If John buys 5 pizzas, how much money will he pay?

    Answer: _____

2. One ice cream cone costs $3.00, and one pastry costs $2.00. Kunal wants to buy 6 ice cream cones and 4 pastries. How much money does he need

    Answer: _____

Write the answer.

3. The cost of 1 T-shirt is $10.00. If Nil buys 3 T-shirts, how much money will he pay?

    Answer: _____

4. 1 wallet costs $9.00, and 1 belt costs $7.00. David wants to buy 2 wallets and 3 belts. How much money does he need?

    Answer: _____

5. Nikita bought 5 bracelets and paid a total of $24.00. She also bought a pair of sunglasses for $28.00. What is the total cost of the bracelets and sunglasses?

    Answer: _____

6. The cost of 1 book is $6.00. If Eli buys 7 books, how much money will he pay?

    Answer: _____

7. Mark bought 20 balls and paid a total of $42.00. He also bought 4 rackets for $32.00. What is the total cost of the balls and rackets?

    Answer: _____

8. 1 pack of nuts costs $8.00, and 1 bottle of juice costs $2.00. Adam wants to buy 4 packs of nuts and 7 bottles of juice. How much money does he need?

    Answer: _____

9. The cost of 1 burger is $3.00. If Ronald buys 10 burgers, how much money will he pay?

    Answer: _____

10. Angela bought 5 lunch boxes and paid a total of $22.00. She also bought 4 sets of cups for $36.00. What is the total cost of the lunch boxes and sets of cups?

    Answer: _____

11. 1 top costs $7.00, and 1 skirt costs $8.00. Simran wants to buy 2 tops and 5 skirts. How much money does she need?

    Answer: _____

12. The cost of 1 pen is $3.00. If Rob buys 6 pens, how much money will he pay?

    Answer: _____

7.5 Profit and Loss (**)

Example 1:

Kavya bought a vacuum cleaner for $190.00 and sold it for $225.00. How much of a profit did she make?

Solution:

The following information is given:

Cost of vacuum cleaner

= $190.00

Selling price of vacuum cleaner

= $225.00

We can solve this problem as follows:

The profit can be calculated by subtracting the cost of the vacuum from the selling price.

Profit = selling price - cost
= $225.00 - $190.00
= $35.00

So Kavya made a profit of $35.00.

Example 2:

Disha spent $81.00 to buy 9 helmets. She sold each helmet for $10.00. How much of a profit did she make?

Solution:

The following information is given:

Number of helmets bought = 9
Cost of helmets = $81.00
Selling price of 1 helmet = $10.00

We can solve this problem as follows:

We need to find the gross amount of money she received from selling the helmets.

Gross amount
= Selling price of 1 helmet
× number of helmets sold
= $10.00 × 9 = $90.00

The profit can be calculated by subtracting the cost of the helmets from the gross amount.

Profit = gross amount - cost
= $90.00 - $81.00
= $9.00

So Disha made a total profit of $9.00.

Write the answer.

1. Maria has spent $64.00 to buy 8 skirts. She sold each skirt for $11.00. How much of a profit did she make?

2. Michael bought a washing machine for $225.00 and sold it for $190.00. How much of a loss did he take?

Answer: _____

Answer: _____

Write the answer.

3. Patricia bought 7 sling bags for $42.00 and sold them for $50.00. How much of a profit did she make?

Answer: _____

4. Nitin bought a mobile phone for $155.00 and sold it for $180.00. How much of a profit did he make?

Answer: _____

5. Georgia has spent $48.00 to buy 6 pounds of cashew nuts. She sold each pound for $5.00. How much of a loss did she take?

Answer: _____

6. Jacob bought 5 markers for $15.00 and sold them for $13.00. How much of a loss did he take?

Answer: _____

7. Carol bought a television for $145.00 and sold it for $172.00. How much of a profit did she make?

Answer: _____

8. Kevin spent $96.00 to buy 8 sets of headphones. He sold each set of headphones for $14.00. How much of a profit did he make?

Answer: _____

9. Diane bought 8 dupattas for $56.00 and sold them for $51.00. How much of a loss did she take?

Answer: _____

10. Edward bought a bike for $840.00 and sold it for $925.00. How much of a profit did he make?

Answer: _____

7.6 Review of Chapter 7 (**)

Write the answer.

1. 1 bag of rice costs $6.00, and 1 bag of sugar costs $9.00. Adam wants to buy 4 bags of rice and 3 bags of sugar. How much money does he need?

   Answer: _____

2. Jacob donates $1,600.00 to a charity, $1,125.00 to a temple, and $650.00 to a youth club every year. How much money does he donate every year?

   Answer: _____

3. If 1 toy costs $9.00, what is the cost of 7 toys?

   Answer: _____

4. Mark bought 6 coffee mugs and paid a total of $12.00. He also bought 3 packages of coffee for $10.00. What is the total cost of the coffee mugs and packages of coffee?

   Answer: _____

5. Paul wants to buy a table that costs $25.00. He also needs to buy a desktop computer that costs $435.00. How much money does he need in total?

   Answer: _____

6. Jessica bought 3 backpacks for $36.00 and sold them for $43.00. How much of a profit did she make?

   Answer: _____

7. Ronald bought a pair of jeans for $27.00 and sold them for $25.00. How much of a loss did he take?

   Answer: _____

8. Sofia wants to buy a dress that costs $62.00. She paid $70.00. How much money did she get back?

   Answer: _____

Write or choose the letter of the answer.

9. Peter bought a remote airplane for $82.00 and sold it for $75.00. How much of a loss did he take?

Answer: _____

10. Rose wants to buy some ice cream cones that cost $35.00. She paid $40.00. How much money did she get back?

Answer: _____

11. Carl wants to buy a bicycle that costs $105.00. He also needs to buy a pair of shoes that costs $45.00. How much money does he need in total?

Answer: _____

12. Diana bought 5 bottles for $4.00 each and 6 glasses for $3.00 each. How much money did she pay in total?

Answer: _____

13. Ellen bought a belt for $25.00 and sold it for $32.00. How much of a profit did she make?

Answer: _____

14. Mrs. Murphy spent $1,400.00 on a home theater system, $450.00 on a mobile phone, and $250.00 on a sari. How much money did she spend in total?

Answer: _____

15. If 3 novels cost $24.00, what is the cost of 1 novel?

Answer: _____

16. Jenny bought 5 lunch boxes and paid a total of $35.00. She also bought some kitchen expenses for $48.00. What is the total cost of the lunch boxes and kitchen expenses?

Answer: _____

8. Work Problems

8.1 Work and Time Concepts – 1 (*)

Example 1:

3 students can solve 18 math questions in a day. How many math questions will 1 student solve in a day?

Solution:

Find the number of math questions that can be solved by 1 student in one day.

Number of math questions that can be solved by 3 students in one day = 18

Number of math questions that can be solved by 1 student in one day

$$= 18 \div 3$$
$$= 6 \text{ math questions}$$

So 1 student can solve 6 math questions in one day.

Note: For a given time, more students can solve more math questions, and fewer students can solve less math questions.

Example 2:

Robert can paint 1 room in 3 hours. How many hours will he take to paint 4 rooms?

Solution:

Find the number of hours required to paint 4 rooms.

Number of hours to paint 1 room = 3 hours

Number of hours to paint 4 rooms

$$= 3 \times 4$$
$$= 12 \text{ hours}$$

Write the answer.

1. It takes 10 seconds to fill a bottle. How long will it take to fill 6 bottles?

    Answer: ____ _____
    unit

2. 4 employees complete 24 assignments in a day. How many assignments can 1 employee complete in a day?

    Answer: ____ _____
    unit

3. Rohan can paint 1 room in 6 hours. How many hours will he take to paint 2 rooms?

    Answer: ____ _____
    unit

4. 4 students can make 24 paper boats in one day. How many paper boats can 1 student make in one day?

    Answer: ____ _____
    unit

5. 1 worker can dig a hole in 2 hours. How many hours will he take to dig 4 holes?

    Answer: ____ _____
    unit

6. 8 carpenters can make 1 door in one day. How many days do they need to make 7 doors?

    Answer: ____ _____
    unit

7. It takes 9 seconds to print a paper. How long will it take to print 6 papers?

    Answer: ____ _____
    unit

8. 1 cook can prepare a dish in 10 minutes. How much time will he take to prepare 5 dishes?

    Answer: ____ _____
    unit

8.2 Work and Time Concepts – 2 (*)

Example 1:

Mr. Parker can design 6 pieces of painting in 12 days. How many days will he take to design 1 piece of painting?

Solution:

We can solve this problem as follows:

Find the time required to design 1 piece of painting.

Time required to design 6 pieces of
painting = 12 days

Time required to design 1 piece of painting
= 12 ÷ 6

= 2 days

So Mr. Parker will take 2 days to design 1 piece of painting.

Example 2:

5 mice can eat 1 slice of cheese in 8 days. If 3 more mice join them, how long will it take them to finish the slice of cheese?

Solution:

We can solve this problem as follows:

Step 1: Find the time it takes 1 mouse to eat a slice of cheese.

Time it takes 5 mice to eat the slice of
cheese = 8 days

Time it takes 1 mouse to eat the slice of
cheese = 8 × 5

= 40 days

Step 2: Find the time taken by 8 mice to eat the slice of cheese.
There will be a total of 8 mice if 3 more mice join them.

Time it takes 1 mouse to eat the slice of
cheese = 40 days

Time it takes 8 mice to eat the slice of
cheese = 40 ÷ 8

= 5 days

So it will take the mice 5 days to finish the slice of cheese.

Write the answer.

1. 5 workers can paint an apartment in 3 days. How many days will it take them to paint 3 apartments?

Answer: _____ _____
unit

2. Kunal can fill 7 buckets in 21 minutes. How many minutes will it take him to fill 1 bucket?

Answer: _____ _____
unit

Write the answer.

3. 4 people can assemble 1 machine in 5 hours. How many hours will it take them to assemble 5 machines?

Answer: ____ _____
unit

4. Rock can write 7 pages in 28 minutes. How many minutes will it take him to write 1 page?

Answer: ____ _____
unit

5. Mr. Johnson can design 5 dresses in 30 days. How many days will it take him to design 1 dress?

Answer: ____ _____
unit

6. 2 workers can dig a hole in 5 hours. How many hours will it take them to dig 2 holes?

Answer: ____ _____
unit

7. Lisa can draw 4 pictures in 28 minutes. How many minutes will it take her to draw 1 picture?

Answer: ____ _____
unit

8. 5 workers can weed a field in 6 hours. How many hours will it take 1 worker to weed the same field?

Answer: ____ _____
unit

9. Disha can run 2 kilometers in 16 minutes. How many minutes will it take her to run 1 kilometer?

Answer: ____ _____
unit

10. 6 students can complete a project in 7 days. How many days will it take them to complete 3 projects?

Answer: ____ _____
unit

11. Andy can read 8 pages in 32 minutes. How many minutes will it take him to read 1 page?

Answer: ____ _____
unit

12. 4 laborers can paint an entrance hall in 2 days. How many days will it take them to paint 3 entrance halls?

Answer: ____ _____
unit

8.3 Work and Time Concepts – 3 (*)

Example 1:

A microwave oven takes 35 minutes to bake 1 cake. How long will it take to bake 4 cakes?

Solution:

Time taken to bake 1 cake = 35 minutes
Time taken to bake 4 cakes = 35 × 4
$$= 140 \text{ minutes}$$

So it will take 140 minutes to bake 4 cakes.

Example 2:

4 carpenters take 12 hours to make a bed. How many hours will it take 1 carpenter to make the same bed?

Solution:

Time taken by 4 carpenters to make a bed
$$= 12 \text{ hours}$$
Time taken by 1 carpenter to make the
same bed = 12 × 4
$$= 48 \text{ hours}$$

So 1 carpenter will take 48 hours to make the same bed.

Example 3:

A group of 24 people can build a 200-foot wall in a week. How many people do we need to build a 500-foot wall in the same amount of time?

Solution:

We can solve this problem as follows:

Step 1: Find the number of people required to build a 1-foot wall in 1 week.

Number of people required to build 200-foot wall = 24

Number of people required to build 1-foot wall = 24 ÷ 200

$$= \frac{24}{200}$$

Step 2: Find the number of people required to build a 500-foot wall in the same amount of time.

Number of people required to build 1-foot wall = $\frac{24}{200}$

Number of people required to build 500-foot wall = $\frac{24}{200} \times 500$

$$= \frac{\overset{12}{\cancel{24}} \times 5}{\cancel{2}_{1}}$$

$$= 12 \times 5$$

$$= 60 \text{ people}$$

So we need 60 people to build a 500-foot wall in the same amount of time.

Write the answer.

1. Lucy takes 6 hours to complete a special task. How long will it take her to do half of the task?

   Answer: _____ _____
   unit

2. A group of boys can fill a tank by using buckets in 3 hours. How long will it take them to fill 3 tanks?

   Answer: _____ _____
   unit

3. Some workers take 1 day to make 4 brick walls. How many brick walls will they make in 2 days?

   Answer: _____ _____
   unit

4. A man takes 9 minutes to run 1 mile. How long will it take him to run 7 miles?

   Answer: _____ _____
   unit

5. Jay takes 9 hours to make a piece of wall art. How long will it take him to make one-third of the wall art?

   Answer: _____ _____
   unit

6. A camera takes 4 seconds to capture a photo. How long will it take to capture 5 photos?

   Answer: _____ _____
   unit

7. Bill takes 6 days to write a storybook. How long will it take him to write two-thirds of the storybook?

   Answer: _____ _____
   unit

8. A juicer takes 5 minutes to produce a glass of juice. How long will it take to produce 7 glasses of juice?

   Answer: _____ _____
   unit

9. Some people take 1 day to paint 3 rooms. How many rooms will they paint in 3 days?

   Answer: _____ _____
   unit

10. A group of workers can make 9 bricks in 1 hour. How long will it take them to make 72 bricks?

    Answer: _____ _____
    unit

8.4 Review of Chapter 8 (*)

Write the answer.

1. 2 science students complete 4 experiments in a day. How many experiments can 1 student complete in a day?

   Answer: _____ _____
                              unit

2. Some workers take 1 hour to make 3 birthday cakes. How many birthday cakes will they make in 4 hours?

   Answer: _____ _____
                              unit

3. Juhi takes 30 minutes to make a piece of art. How long will it take her to make one-third of the piece of art?

   Answer: _____ _____
                              unit

4. It takes 10 seconds to fill 1 juice glass. How long will it take to fill 4 juice glasses?

   Answer: _____ _____
                              unit

5. A water tap takes 4 hours to empty a tank. How long will it take to empty 3 tanks?

   Answer: _____ _____
                              unit

6. 1 carpenter can make a door in 2 days. How many days will it take him to make 6 doors?

   Answer: _____ _____
                              unit

7. 8 students can complete a project in 5 days. How many days will it take them to complete 4 projects?

   Answer: _____ _____
                              unit

8. Jack can fill 7 water bottles in 56 seconds. How many seconds will it take him to fill 1 water bottle?

   Answer: _____ _____
                              unit

Write the answer.

9. 4 painters can paint an apartment in 5 days. How many days will it take them to paint 3 apartments?

Answer: ____ _____
unit

10. David can fill 8 buckets in 32 minutes. How many minutes will it take him to fill 1 bucket?

Answer: ____ _____
unit

11. Some engineers take 1 day to design 5 cars. How many cars will they design in 2 days?

Answer: ____ _____
unit

12. 4 robots can assemble 1 machine in 6 hours. How many hours will it take them to assemble 5 machines?

Answer: ____ _____
unit

13. It takes 2 hours to watch a movie. How long will it take to watch 3 movies?

Answer: ____ _____
unit

14. 1 tiger can run 1 kilometer in 4 minutes. How many minutes will it take to run 3 kilometers?

Answer: ____ _____
unit

15. Bob takes 20 minutes to make a special dish. How long will it take him to make half of the dish?

Answer: ____ _____
unit

16. A group of people can fill a water tank using buckets in 2 hours. How long will it take them to fill 4 water tanks?

Answer: ____ _____
unit

9. Mixture Problems

9.1 Mixture Problems with Objects (*)

Example 1:

A box has 25 green marbles and 15 white marbles. If we mix the marbles, what will be the total number of marbles in the box?

Solution:

The following information is given:

Number of green marbles = 25

Number of white marbles = 15

You can find the answer as shown below:

Total marbles

= Number of green marbles
   + Number of white marbles

= 25 + 15 = 40 marbles

So there are a total of 40 marbles in the box.

Example 2:

Jar A has 10 mango candies and 15 orange candies. Jar B has 8 mango candies and 11 orange candies. How many more candies are there in Jar A than in Jar B?

Solution:

The following information is given:

Number of mango candies in Jar A = 10
Number of orange candies in Jar A = 15
Number of mango candies in Jar B = 8
Number of orange candies in Jar B = 11

Step 1: Find the total number of candies in Jar A.

= Number of mango candies
   + Number of orange candies

= 10 + 15 = 25 candies

Step 2: Find the total number of candies in Jar B.

= Number of mango candies
   + Number of orange candies

= 8 + 11 = 19 candies

Step 3: Find the number of candies in Jar A that exceed the number of candies in Jar B.

= total candies in Jar A
   − total candies in Jar B

= 25 − 19 = 6 candies

So there are 6 more candies in Jar A than in Jar B.

Write the answer.

1. A packet has 56 red balloons and 48 blue balloons. What is the total number of balloons in the packet?

    Answer: _____ _____
    unit

2. Bag 1 has 65 paper boats, and Bag B has 52 paper boats. How many paper boats do we need to add to Bag B so that the number of paper boats is the same in both the bags?

    Answer: _____ _____
    unit

3. A jar has 32 marbles. How many marbles do you need to add to the jar so that the number of marbles will be 40?

    Answer: _____ _____
    unit

4. Box A has 21 shirts and 13 pairs of pants. Box B has 25 shirts and 15 pairs of pants. How many more items are there in Box B than in Box A?

    Answer: _____ _____
    unit

5. A box has 35 crayons. How many crayons do you need to add to the box so that the number of crayons will be 50?

    Answer: _____ _____
    unit

6. A packet has 15 black sketches and 20 red sketches. What is the total number of sketches in the packet?

    Answer: _____ _____
    unit

7. Basket A has 38 oranges, and Basket B has 45 oranges. How many oranges do we need to add to Basket A so that the number of oranges is the same in both the baskets?

    Answer: _____ _____
    unit

8. A packet has 42 stickers. How many stickers do you need to add to the packet so that the number of stickers will be 60?

    Answer: _____ _____
    unit

9.2 Mixture Problems with Solutions (*)

Example 1:

Jar A has 45 milliliters (ml) of honey and 30 milliliters (ml) of water. Jar B has 25 ml more honey than Jar A. What is the amount of honey in Jar B?

Solution:

The following information is given:

Amount of honey in Jar A = 45 ml

Amount of water in Jar A = 30 ml

Amount of honey in Jar B = 25 ml more than Jar A

We can solve this problem as follows:

Amount of honey in Jar B
  = Amount of honey in jar A + 25 ml
  = 45 ml + 25 ml
  = 70 ml

So there is 70 milliliters of honey in Jar B.

Example 2:

Can A has 400 milliliters (ml) of milk and 100 milliliters (ml) of water. Can B has 100 ml more milk than Can A and 300 ml of water. If we mix the contents of both of the cans, what is the total amount of solution?

Solution:

The following information is given:

Amount of milk in Can A = 400 ml

Amount of water in Can A = 100 ml

Amount of milk in Can B = 100 ml more than can A

Amount of water in Can B = 300 ml

We can solve this problem as follows:

Step 1: Find the amount of milk in Can B.

Amount of milk in Can B
  = Amount of milk in Can A + 100 ml
  = 400 ml + 100 ml
  = 500 ml

Step 2: Find the total solution in Can A.

Amount of solution in Can A
  = Amount of milk in Can A
       + Amount of water in Can A
  = 400 ml + 100 ml
  = 500 ml

Step 3: Find the total solution in Can B.

Amount of solution in Can B
  = Amount of milk in Can B
       + Amount of water in Can B
  = 500 ml + 300 ml
  = 800 ml

Step 4: Find the total solution in both cans.

Amount of solution in both the cans
  = Amount of solution in Can A
       + Amount of solution in Can B
  = 500 ml + 800 ml
  = 1,300 ml

So the total amount of solution is 1,300 milliliters.

Write the answer.

1. A 75-milliliter bottle of shampoo has 25 milliliters of water. What is the quantity of shampoo in the bottle?

Answer: _____ _____
unit

2. Jar A has 350 milliliters of petrol and 200 milliliters of kerosene. Jar B has 200 milliliters more petrol than Jar A and 450 milliliters of kerosene. If we mix the contents of both the jars, what is the total amount of solution?

Answer: _____ _____
unit

3. Bottle A has 80 milliliters of water and 20 milliliters of alcohol. Bottle B has 65 milliliters of water and 35 milliliters of alcohol. If we mix the contents of both of the bottles, what is the total amount of water in the solution?

Answer: _____ _____
unit

4. Bottle 1 has 50 milliliters of water and 25 milliliters of alcohol. Bottle 2 has 35 milliliters more water than Bottle 1. What is the amount of water in Bottle 2?

Answer: _____ _____
unit

5. Bowl 1 has 400 milliliters of milk and 100 milliliters of water. Bowl 2 has 300 milliliters of milk and 250 milliliters of water. If we mix the contents of both of the bowls, what is the total amount of milk in the solution?

Answer: _____ _____
unit

6. A 150-milliliter bottle of honey has 50 milliliters of water. What is the quantity of honey in the bottle?

Answer: _____ _____
unit

7. Bottle A has 75 milliliters of shampoo and 25 milliliters of alcohol. Bottle B has 20 milliliters more shampoo than Bottle A. What is the amount of shampoo in Bottle B?

Answer: _____ _____
unit

8. Bucket 1 has 500 milliliters of water and 50 milliliters of detergent liquid. Bucket 2 has 75 milliliters more water than Bucket 1 and 60 milliliters of detergent liquid. If we mix the contents of both of the buckets, what is the total amount of solution?

Answer: _____ _____
unit

9.3 Review of Mixture Problems (*)

Write the answer.

1. A jar has 40 red marbles and 32 blue marbles. If we mix the marbles, what will be the total number of marbles in the jar?

   Answer: _____ _____
   unit

2. Jar A has 200 milliliters of juice and 100 milliliters of water. Jar B has 50 milliliters more huice than Jar A and 200 milliliters of water. If we mix the contents of both of the jars, what is the total amount of solution?

   Answer: _____ _____
   unit

3. Basket 1 has 20 bananas and 15 apples. Basket 2 has 13 bananas and 20 apples. How many more fruits are there in Basket 1 than in Basket 2?

   Answer: _____ _____
   unit

4. A 500-milliliter bottle of honey has 100 milliliters of water. What is the quantity of honey in the bottle?

   Answer: _____ _____
   unit

5. Can A has 20 liters of milk and 2 liters of water. Can B has 17 liters of milk and 3 liters of water. If we mix the contents of both of the bottles, what is the total amount of milk in the solution?

   Answer: _____ _____
   unit

6. Bag A has 50 toys, and Bag B has 34 toys. How many toys do we need to add to Bag B so that the number of toys is the same in both the bags?

   Answer: _____ _____
   unit

7. A cupboard has 12 shirts. How many shirts do you need to add to the cupboard so that the number of shirts will be 25?

   Answer: _____ _____
   unit

8. Bottle 1 has 150 milliliters of water and 40 milliliters of alcohol. Bottle 2 has 25 milliliters more water than Bottle 1. What is the amount of water in Bottle 2?

   Answer: _____ _____
   unit

Write the answer.

9. Bag A has 50 pounds of rice, and Bag B has 60 pounds of rice. How much rice do we need to add to Bag A so that the amount of rice is the same in both the bags?

Answer: ____ _____
unit

10. Bottle 1 has 500 milliliters of juice and 100 milliliters of alcohol. Bottle 2 has 150 milliliters more juice than Bottle 1 and 200 milliliters of alcohol. If we mix the contents of both of the bottles, what is the total amount of solution?

Answer: ____ _____
unit

11. A 150-milliliter bottle of cough syrup has 40 milliliters of alcohol. What is the quantity of cough syrup in the bottle?

Answer: ____ _____
unit

12. Jar A has 600 milliliters of kerosene and 150 milliliters of oil. Jar B has 175 milliliters more kerosene than Jar A. What is the amount of kerosene in Jar B?

Answer: ____ _____
unit

13. Packet A has 70 mango candies and 30 strawberry candies. Packet B has 52 mango candies and 40 strawberry candies. How many more candies are there in Packet A than in Packet B?

Answer: ____ _____
unit

14. An aquarium has 30 red fish and 28 goldfish. If we mix the fish, what will be the total number of fish in the aquarium?

Answer: ____ _____
unit

15. Drum 1 has 300 liters of alcohol and 60 liters of water. Drum 2 has 280 liters of alcohol and 40 liters of water. If we mix the contents of both of the drums, what is the total amount of alcohol in the solution?

Answer: ____ _____
unit

16. A basket has 20 bottles. How many bottles do you need to add to the basket so that the number of bottles will be 35?

Answer: ____ _____
unit

9.4 Filling or Emptying a Tank – 1 (*)

Example 1:

A tap can fill a bucket in 4 minutes. How long will it take the tap to fill half of the bucket?

Solution:

You can consider the whole bucket to be 1 and find the answer as shown below:

Time the tap takes to fill the whole (1) bucket = 4 minutes

Time the tap takes to fill half of the bucket

$$= 4 \div 2$$
$$= 2 \text{ minutes}$$

The tap will take 2 minutes to fill half of the bucket.

Example 2:

Pump A can fill a container in 3 hours. If Pump B takes 2 hours more than Pump A to fill the same container, how long will it take Pump B to fill the container?

Solution:

Time Pump A takes to fill the whole container = 3 hours

Pump B takes 2 hours more than Pump A to fill the same container.

Time taken by Pump B

$$= \text{(Time taken by Pump A)} + 2$$
$$= 3 \text{ hours} + 2$$
$$= 5 \text{ hours}$$

Pump B takes 5 hours to fill the same container.

Write the answer.

1. A pipe can empty 20 liters of kerosene from a drum in 4 minutes. If the pipe is opened for 2 minutes, how much kerosene can be emptied?

Answer: _____ _____
unit

2. Pipe 1 can empty a water tank in 40 minutes. Pipe 2 can empty the same water tank 10 minutes faster than Pipe 1. How long will it take Pipe 2 to empty the tank?

Answer: _____ _____
unit

3. In a milk factory, a pipe can fill the milk container in 10 minutes. How long will it take to fill half of the milk container?

Answer: _____ _____
unit

4. Water Tap A can empty a tank in 35 minutes. Water Tap B takes 20 minutes longer to empty a tank than Water Tap A. How long will it take Water Tap B to empty a tank?

Answer: _____ _____
unit

Write the answer.

5. A pump can fill a pool in 30 minutes. How long will it take to fill half of the pool?

    Answer: _____ _____
    unit

6. A water pipe can empty a tank in 40 minutes. How long will it take to empty half of the tank?

    Answer: _____ _____
    unit

7. Pump 1 can fill a container in 30 minutes. Pump 2 takes 15 minutes longer than Pump 1 to fill the same container. How long will it take Pump 2 to fill the container?

    Answer: _____ _____
    unit

8. Tap A can empty a glass container in 25 minutes. Tap B takes 10 minutes longer than Tap A to empty the same container. How long will it take Tap B to empty the glass container?

    Answer: _____ _____
    unit

9. Pump 1 can fill a drum in 20 minutes. Pump 2 takes 5 minutes longer than Pump 1 to fill the same drum. How long will it take Pump 2 to fill the drum?

    Answer: _____ _____
    unit

10. A pipe can empty a drum in 20 minutes. How long will it take to empty half of the drum?

    Answer: _____ _____
    unit

11. Tap A can fill a container in 45 minutes. Tap B can fill the same container in 20 minutes less than Tap A. How long will it take Tap B to fill the container?

    Answer: _____ _____
    unit

12. Water Tap 1 can fill a tank in 3 hours. Water Tap 2 can fill the same tank in 2 hours less than Water Tap 1. How long will it take Water Tap 2 to fill the tank?

    Answer: _____ _____
    unit

9.5 Filling or Emptying a Tank – 2 (*)

Example 1:

A small pool has 4 pipes to empty the water. Each pipe can empty 8 liters of water in 1 minute. If all the pipes are opened, how much water is emptied in 1 minute?

Solution:

The following information is given:

Number of pipes in a small pool = 4

Amount of water each pipe can empty in 1 minute = 8 liters

Find the amount of water that can be emptied by all pipes in 1 minute.

Amount of water emptied in 1 minute
= number of pipes × water emptied by 1 pipe in 1 minute
= 4 × 8 liters = 32 liters

32 liters of water are emptied in 1 minute if all the pipes are opened.

Example 2:

An oil tank has a capacity of 980 liters. The tank has 750 liters of oil. How much more oil is required to fill the tank?

Solution:

The following information is given:

Capacity of an oil tank = 980 liters

Amount of oil in the tank = 750 liters

We can solve this problem as follows:

Find the additional amount of oil required to fill the tank.

Additional amount of oil required

$$= 980 - 750$$

$$= 230 \text{ liters}$$

So 230 additional liters of oil are required to fill the tank.

Write the answer.

1. Pipe 1 can fill a container in 45 minutes. Pipe 2 takes 15 minutes longer than Pipe 1 to fill the same container. How long will it take Pipe 2 to fill the container?

    Answer: _____ _____
    unit

2. A tank has 5 pipes to empty the water. Each pipe can empty 3 liters of water in 1 minute. If all the pipes are opened, how much water is emptied in 1 minute?

    Answer: _____ _____
    unit

Write the answer.

3. A tank has 3 pipes to empty the water. Each pipe can empty 7 liters of water in 1 minute. If all the pipes are opened, how much water is emptied in 1 minute?

Answer: ____ _____
                  unit

4. A container is filled with milk. Because of a small leak at the bottom, 24 milliliters of milk leak out in 3 minutes. How much milk will leak out in 6 minutes?

Answer: ____ _____
                  unit

5. Tap 1 can fill a swimming pool in 14 hours. Tap 2 takes 4 hours more than Tap 1 to fill the swimming pool. How long will it take Tap 2 to fill the pool?

Answer: ____ _____
                  unit

6. A container has 4 pipes to fill water. Each pipe can fill 5 liters in 1 minute. If all the pipes are opened to fill water at the same time, how much water is filled in 1 minute?

Answer: ____ _____
                  unit

7. An oil tank has a capacity of 850 liters. The tank has 630 liters of oil. How much more oil is required to fill the tank?

Answer: ____ _____
                  unit

8. Pipe 1 can empty a container in 40 minutes. Pipe 2 can empty the same container in 16 minutes less than Pipe 1. How long will it take Pipe 2 to empty the container?

Answer: ____ _____
                  unit

9. Oil Tank 1 has a capacity of 650 liters. Oil Tank 2 has a capacity 200 liters more than Oil Tank 1. What is the capacity of Oil Tank 2?

Answer: ____ _____
                  unit

10. A tank has 6 pipes to empty water. Each pipe can empty 4 liters of water in 1 minute. If all the pipes are opened, how much water is emptied in 1 minute?

Answer: ____ _____
                  unit

Quiz

1. 1 ball costs $2.00. Which math sentence will you use to find the cost of 25 balls?
   (a) 25 × 2.00
   (b) 2.00 + 25
   (c) 25 ÷ 2.00
   (d) All of the above

   Answer: _____

2. What is the operation key word(s) in the following problem?

   What is the product of 3 and 5?

   Answer: _____

3. The cost of 1 shampoo bottle is $4.00. Adam bought 4 of them and gave $20.00 to the cashier. How much money will the cashier return?

   Answer: _____

4. Review the question given below and choose the best answer about the available information.

   Peter wants to buy 1 notebook. How much money does he need to pay for the notebook?

   (a) Too much information
   (b) Too little information
   (c) The right amount of information

   Answer: _____

5. Angela spent $42.00 at a shopping mall. She bought 6 T-shirts that cost the same amount. What is the cost of each T-shirt?

   Answer: _____

6. Bob and Maria can paint a room in 3 hours. If Bob wants to paint alone, how many hours will he take to paint the room?

   Answer: ____ _____
                        unit

7. Max writes 32 pages in 4 days. How many pages can he write in 1 day?

   Answer: ____ _____
                        unit

8. If you replace the ones digit by the double of the hundreds digit in the number 258, what is the new number?

   Answer: _____

9. Write 579 in expanded form and find the missing number in the following math sentence:

   579 = 500 + _____ + 9

   Answer: _____

10. Simran is currently 6 years older than Loni. If Loni is 15 years old now, how old is Simran?

Answer: _____

11. Nelson's age is one-half of his brother's age. His brother's age is 10. What is Nelson's age?

Answer: _____

12. A cultural program started at 5:00 p.m. and ended at 8:00 p.m. How long was the program?

Answer: _____ _____
unit

13. A cyclist has to ride 40 miles. How long will it take him if he can ride 8 miles in 1 hour?

Answer: _____ _____
unit

14. If April 14 is on Sunday, which day will it be on April 18?

Answer: _____

15. Nancy wants to buy a scooter that costs $650.00. She paid $700.00. How much money did she get back?

Answer: _____

16. Alan bought 4 flower vases for $10.00 each and 5 paintings for $8.00 each. How much money did he pay in total?

Answer: _____

17. Some workers take 1 hour to prepare 4 dishes. How many dishes will they prepare in 3 hours?

Answer: _____ _____
unit

18. A camera can capture 10 photos in 10 seconds. How many seconds will it take to capture 1 photo?

Answer: _____ _____
unit

19. Bag 1 has 48 notes, and Bag 2 has 25 notes. How many notes do we need to add to Bag 2 so that the number of notes is the same in both bags?

Answer: _____ _____
unit

20. A tap can fill a tank in 3 hours. How long will it take to fill 4 tanks?

Answer: _____ _____
unit

21. Pipe 1 can empty a water tank in 50 minutes. Pipe 2 takes 12 minutes longer than Pipe 1 to empty the same water tank. How long will it take Pipe 2 to empty the tank?

Answer: _____ _____
unit

www.ingramcontent.com/pod-product-compliance
Lightning Source LLC
Chambersburg PA
CBHW060008210526
45170CB00017B/2074